Computer Graphics for

LANDSCAPE ARCHITECTS

An Introduction

COMPUTER GRAPHICS FOR

LANDSCAPE ARCHITECTS

AN INTRODUCTION

ASHLEY CALABRIA
AND
JOSE BUITRAGO

DELMAR
CENGAGE Learning

Australia • Brazil • Japan • Korea • Mexico • Singapore • Spain • United Kingdom • United States

DELMAR
CENGAGE Learning™

Computer Graphics for Landscape Architects: An Introduction
Ashley Calabria and Jose R. Buitrago

Vice President, Career and Professional Editorial:
Dave Garza

Director of Learning Solutions:
Matthew Kane

Acquisitions Editor:
David Rosenbaum

Managing Editor:
Marah Bellegarde

Product Manager:
Christina Gifford

Editorial Assistant:
Scott Royael

Vice President, Career and Professional Marketing:
Jennifer McAvey

Marketing Director:
Deborah Yarnell

Marketing Manager:
Gerard McAvey

Marketing Coordinator:
Jonathan Sheehan

Production Director:
Wendy Troeger

Production Manager:
Mark Bernard

Content Project Manager:
Katie Wachtl

Art Director:
Dave Arsenault

Technology Project Manager:
Sandy Charette

Production Technology Analyst:
Thomas Stover

For product information and technology assistance, contact us at
Professional & Career Group Customer Support, 1-800-648-7450

For permission to use material from this text or product,
submit all requests online at **cengage.com/permissions**
Further permissions questions can be e-mailed to
permissionrequest@cengage.com

Library of Congress Control Number: 2008925687

ISBN-13: 978-1-4180-6525-6

ISBN-10: 1-4180-6525-0

Delmar
5 Maxwell Drive
Clifton Park, NY 12065-2919
USA

Cengage Learning is a leading provider of customized learning solutions with office locations around the globe, including Singapore, the United Kingdom, Australia, Mexico, Brazil, and Japan. Locate your local office at: **international.cengage.com/region**

Cengage Learning products are represented in Canada by Nelson Education, Ltd.

For your lifelong learning solutions, visit **delmar.cengage.com**

Visit our corporate website at **cengage.com**

Notice to the Reader

Publisher does not warrant or guarantee any of the products described herein or perform any independent analysis in connection with any of the product information contained herein. Publisher does not assume, and expressly disclaims, any obligation to obtain and include information other than that provided to it by the manufacturer. The reader is expressly warned to consider and adopt all safety precautions that might be indicated by the activities described herein and to avoid all potential hazards. By following the instructions contained herein, the reader willingly assumes all risks in connection with such instructions. The publisher makes no representations or warranties of any kind, including but not limited to, the warranties of fitness for particular purpose or merchantability, nor are any such representations implied with respect to the material set forth herein, and the publisher takes no responsibility with respect to such material. The publisher shall not be liable for any special, consequential, or exemplary damages resulting, in whole or part, from the readers' use of, or reliance upon, this material.

Printed in Canada
1 2 3 4 5 6 7 12 11 10 09 08

Contents

Chapter 2   SketchUp / 79

by Professor Ashley Calabria / 79

Chapter **3** From AutoCAD to Adobe Photoshop CS2 Rendering / 123

by Professor Jose R. Buitrago / 123

Chapter 5 InDesign   /   219

by Professor Ashley Calabria   /   219

Chapter 6 Program Interchange and Student Project Examples / 237

by Professors Ashley Calabria and Jose R. Buitrago / 237

Dedication

This book is dedicated to
our students at the University of Georgia
School of Environmental Design,
whose creativity and ambition has inspired us over the years.

Preface

 ## By Professor Ashley Calabria

In my thirteen years, our department has taught over a dozen different computer programs through a variety of different classes. It became apparent that our department and the profession itself were experiencing a paradigm shift from hand to computer graphics, and between graphic programs. Out of curiosity and with the hope of keeping our computer graphics curriculum forward-looking, I started surveying our interns to find out how much hand graphics and computer graphics they were using, as well as what programs they were using during their internships. Every summer we place 60 to 80 interns in a variety of landscape-architecture-related fields, primarily throughout, but not limited to, the southeastern United States. The results are proving quite interesting.

 ## By Professor Jose R. Buitrago

There is a saying that "two heads think better than one," and this book is a clear example of this almost universal concept. The authors of this book, after several years of struggling to find a single book that fit the entire scope of the computer graphics course they were teaching to both undergraduate and graduate students of landscape architecture at the University of Georgia, decided over a "latte gathering" to take on the challenge themselves, and combined their efforts to write this book. Utilizing our own notes, experience, and materials that we generated for these classes, we wrote this book with the intent of harnessing that information into a condensed, introductory, and easy-to-follow tutorial format for both landscape architecture students and professionals who are interested in using more technology but who never had the time or chance to learn it.

The question of where to start was the initial research question that inspired Assistant Professor Ashley Calabria in her academic investigation into tracking practitioners' shift from hand to computer graphics, as computer graphics programs increasingly took hold in landscape architecture firms over the last several years. Calabria's research methodology led to the development of a survey that seeks to identify the computer graphics programs most commonly used by students during their required summer professional internship. This survey has also helped us evolve our computer graphics courses and inform students of the trends and skills in the current job market.

Hand and Computer Graphics

The term "virtual office" has been circulating for several years, and since Professor Buitrago and I teach both hand and computer graphics, we are interested in the practicality of this. The surveys are unraveling a compelling story that seems to question this paradigm shift from hand to computer graphics as a one-way transition.

The first survey was sent out in 2004 to University of Georgia, School of Environmental Design (UGA SED) interns. Sixty-three surveys were returned. The first question we wanted to address was how much hand or computer graphics were being used in anticipation of potential hand and computer course curriculum shifts.

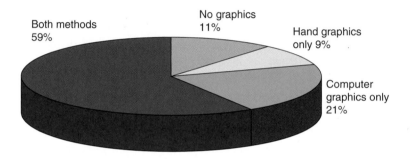

The 2004 results, shown in the preceding figure, did not provide such a surprising statistic, but over time I assumed that the number of students who used only computer graphics would increase, with numbers drawing from the other categories of hand graphics, no graphics and both methods. However, in the 2006 survey, 65 were returned, showing an increase in the number of students using both methods and a decrease in students using solely computer graphics and, thankfully, a decrease in those using no graphics in their internships.

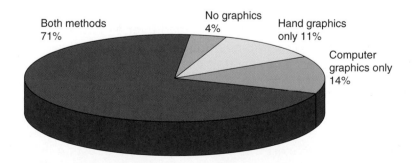

This statistical data demonstrate that both methods still play a valuable role in the field of landscape architecture. Evaluating why is the topic for another book, but when asked, firms seem to express a nonquantitative value on how design is registered and carried out via hand or computer. This initiated a survey in 2006 that was sent to firms hiring our interns. They were asked to identify the top skills they wanted to see in student resumes and portfolios. In the 26 responses, the top four skill sets were

> Clean/organized layout
>
> Computer graphics
>
> Hand graphics
>
> Conceptual graphics and writing skills (tied)

The fact that firms still rank hand graphics and other forms of traditional communication so highly, is undeniably evident in both surveys and personal interviews. As professors of hand graphics as well, we both use this data to demonstrate to students that there is no definitive answer on using one method over the other for the for the different levels of graphic communication needed. Both methods should be valued for their benefits to the different forms of graphic representation and communication throughout the design process.

Computer Programs for Landscape Architects

Computer applications are no exception to the shift in graphic communication in design fields. So how were the programs for this book chosen? Returning to the intern surveys of 2004 and 2006, we can see some dramatic changes that occurred in just two years.

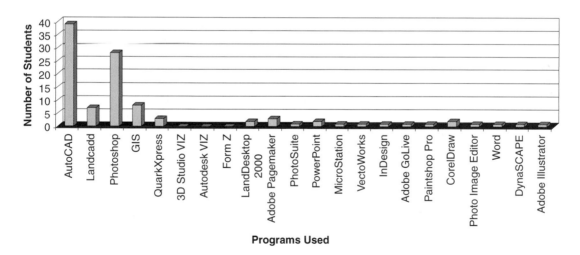

In 2004 AutoCAD was the most commonly used program, and Photoshop came in second with just over half the students using it in their internships. A few stated using GIS and Landcadd as well as Quark Xpress and Adobe Pagemaker.

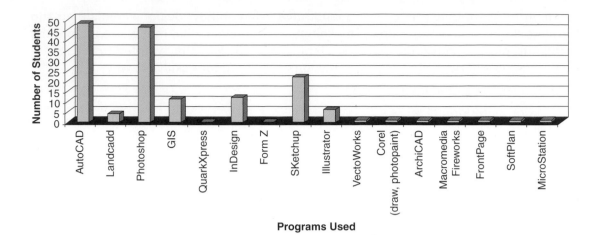

In 2006 the statistical data identified that interns were using nearly as much Photoshop as AutoCAD. And most surprisingly, SketchUp was the third most widely used program in the 2006 results, although in 2004 it was not even listed under the "other" category. Whereas AutoCAD, Photoshop, and SketchUp ranked as most widely used by SED interns, employers in a similar question (not shown) ranked using AutoCAD and Photoshop equally, PowerPoint, and then SketchUp.

The most recent survey, sent out to firms in the summer of 2007, reinforces the use of the programs selected for this book. Thirty-six firms responded, stating that more offices used Photoshop than AutoCAD, with PowerPoint coming in third and SketchUp fourth.

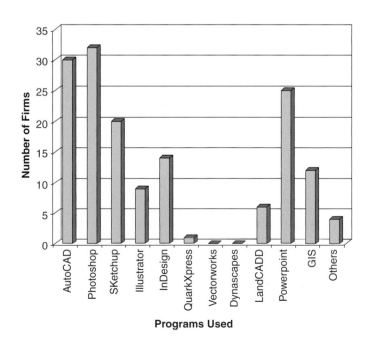

The most common project types created by computer graphics programs were for drafting, presentation drawings, digital imagery, rendering, and creating section-elevations, respectively.

This information has been well received at conferences, published in proceedings, and written about in articles. But equally important is providing this information to students, encouraging opportunities to develop a variety of communication skills and to open a dialogue between academia and the profession during this time of graphic change.

This research will continue as a way to capture the transition our field is experiencing, to document the trends, and to investigate new technologies as they advance. Future editions of this book will reflect those changes. We hope you find this information and book helpful. Remember

- Take breaks when working on the computer.
- Be patient with yourself as you learn these new skills.
- Don't worry about making mistakes.
- Save often.
- And try to have fun with it!

About the Authors

 ## Ashley Calabria

Ashley Calabria is an Assistant Professor with the University of Georgia, School of Environmental Design. She completed her undergraduate degree from the University of North Carolina, Asheville and holds an MLA from the University of Georgia. Professor Calabria is an associate editor of digital information for *Landscape Architect and Specifier News* and has authored a number of peer reviewed conference proceeding articles discussing the use of computer programs in the landscape design field. She is a member of the American Society of Landscape Architects and The Design Communication Association.

 ## Jose R. Buitrago

Jose R. Buitrago is an Assistant Professor with the University of Georgia, School of Environmental Design. Professor Buitrago completed his undergraduate degree from The Pennsylvania State University and obtained his MLA from Harvard University, Graduate School of Design. Professor Buitrago has authored a number of articles discussing the use of computer graphics in landscape architecture. He is a member of the Northeast Georgia Section of the American Society of Landscape Architecture and a member of *Men's Health Living* magazine board of expert advisors.

Acknowledgments

 ## Ashley Calabria

Thank you, mom and dad, for raising me with love and support. For my wonderful husband, Gil, and my magnificent kids, Lorenzo and Chloe, thank you for being patient and for always making me smile when I come home. Love.

 ## Jose R. Buitrago

I wish to express my sincere thanks to my colleagues, students, and staff at the University of Georgia's School of Environmental Design for their unconditional support and accommodations in the writing of this book. Many thanks to Marsha Grizzle, who never said no to my short notice grammar spell check request. To my colleague and friend, Ashley Calabria, who despite her hectic schedule, joined me in the pursuit of this venture. Last but not the least, to my family, I will be forever grateful for your unconditional love, patience, and support.

Additionally, the authors and Delmar Cengage Learning would like to thank the following individuals for their time and professional expertise in reviewing the manuscripts:

David Hopman
The University of Texas at Arlington
Arlington, TX

Elizabeth Mogen
Colorado State University
Fort Collins, CO

Sean P. Michael
Washington State University
Pullman, WA

Phillip S. Waite
Washington State University
Pullman, WA

AutoCAD

By Professor Jose R. Buitrago

CHAPTER 1

CHAPTER OBJECTIVE

This chapter introduces you to the two-dimensional drafting software known as **AutoCAD**. The first part will give a general overview of the digital drafting tools and will be followed by detailed instructions on how to establish the basic settings of a drawing. By the end of this chapter you should be able to draft a plan-view drawing, and prepare the document for a hard copy.

Introduction to AutoCAD

In 1982, a group of 16 programmer friends founded a company by the name of **AutoDesk**, in the city of San Rafael, California. With a limited investment, these visionaries' programmers created one of the most popular computer-generated drafting softwares in the world. They called this software CADD, which stands for computer-aided design and drafting. The main objective of this software was to aid designers who needed precision drafting with capabilities for design development. At the time of the creation of this software, personal desktop computers were in their evolutionary infancy. CADD was regarded as a secondary tool for landscape architects because of its inability to deal with curvilinear elements, rudimentary nature, and less artistic look. Today, with the giant leap in microprocessor technology and desktop computers, CADD is no longer considered a precision drafting tool but a rendering software as well.

The author's research of the most commonly used software, and their common daily application in a landscape architecture professional practice, set and led the format of this book. Even though the reader may be tempted to commence drafting right away, it is important to understand all the basic notions, tools, icons, and screen formats, and the process of setting a CAD drawing, before beginning the drafting process. Later in this chapter, the authors will guide the reader step by step on how to draft a first drawing. Also, the author's research supported the exploration of CAD as a nontraditional rendering tool (rather than technical drafting), due to the demand by landscape architecture firms and practitioners who require potential employees and upcoming landscape architecture graduates to have this knowledge, skill, and expertise. This chapter will guide you through a learning exploration of CAD (shorter version for CADD) beyond its capabilities as a precision drafting tool, and in the orderly format that is commonly used in landscape architecture professional practice. Let us get started.

AutoCAD Screen and Preferences

The AutoCAD default screen

The first step is to launch the latest version of **AutoCAD (ACAD, CADD,** or **CAD)**. Simply double-click the left button of your mouse on the AutoCAD icon on your desktop. Also, you can launch the software by browsing under the Start button located on the bottom toolbar of your desktop window, then open **All Programs**, scroll to **AutoDesk** (My) folder, next **AutoCAD 2007**, and finally click on the **AutoCAD 2007** icon. Figure 1-1 illustrates the default AutoCAD window that will automatically open. Depending on your CAD default settings, your screen might look a little different from figure 1-1, but in general will look the same.

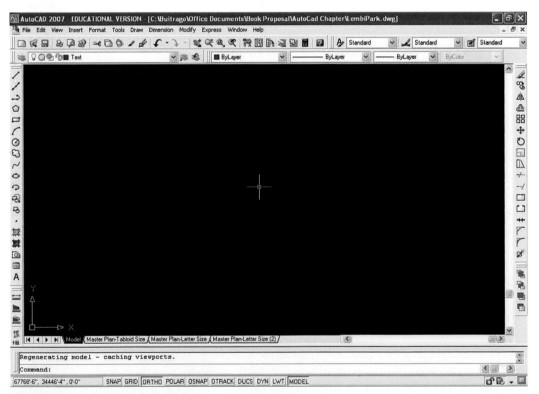

FIGURE 1–1 The AutoCAD default screen.

The next step is to familiarize yourself with the standard AutoCAD window. Hold the cursor arrow (using the mouse) on top of each icon/button for a few seconds; a yellow tag window will pop up (open), with the name of the tool and the key typo command.

Later in this chapter, we will explain how to use the key typo command. Make sure to take the time to learn all the different names of the tool icon/buttons as we will be referring to them by their names (figure 1-2).

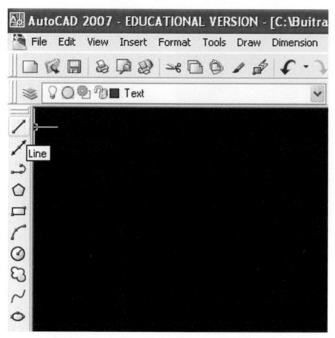

FIGURE 1–2 A yellow tag window will pop up for the Line tool icon.

On top of the screen you will see the default title bar showing the file path where the drawing is located and the name/title of the drawing. Underneath the title bar you will see the followings words: File , Edit , Insert , Format , Tools , Draw , Dimension , Modify , Window , etc. See figure 1-3. This is the menu bar. Below the menu bar is the standard toolbar showing several CAD icon tools. At the bottom of the screen, you will see a window that reads "Regenerating model, AutoCAD utilities loaded, and Command." This is called the Command Prompt window (figure 1-4). Above this you will find the Model, Layout 1, and Layout 2 tags. Below the Command Prompt window are the Snap, Grid, Ortho, Polar, Osnap, Otrack, LWT, and Paper tags. Depending on the default settings of your CAD software you may also see several tool icons on either side of your screen. These are called the left or right toolbars.

FIGURE 1–3 The menu bar and standard toolbar.

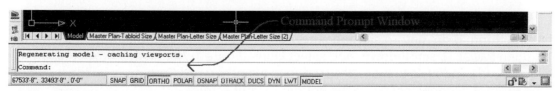

FIGURE 1–4 The Command Prompt window.

Place the arrow on top of the Layout 1 tag and right-click the mouse; this will launch a pop-up menu window where the option Rename tag among others is listed (figure 1-5). Further explanations of theses options will be covered later in this chapter.

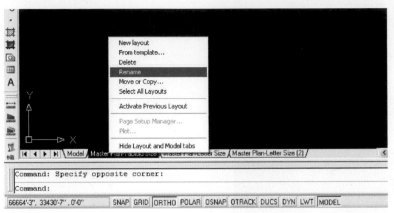

FIGURE 1–5 The Rename tag in the pop-up menu window.

Model space and paper space

The center of the screen shows a black screen with nothing on it. This is called model space. When your screen is in the model space mode, the Model tag on the top of the Command Prompt window is highlighted in black. If you click the next tag label, Layout 1 (or whatever name you have given the tag), it will open the screen into what is called paper space (figure 1-6). CAD drafting works in two modes: model space and paper space. In theory, model space is like a base sheet of paper that is placed underneath a trace paper, which is the paper space. In order to see model space, you must cut a window or hole in paper space. The reasoning behind this arrangement is that such a window (called **Viewports**) allows you to either show the entire domain of model space or portions of it. Figure 1-6 shows the viewport as a solid rectangular line; the dashed line illustrates the printing margin limits of the default printer, and the white area represents the limits of the paper. This format facilitates your drafting since, in model space, there is no scale limit. Whatever you draw in model space, regardless of the dimension, the virtual environment will expand proportionately to accommodate it. On the other hand, paper space is limited to the size of your paper dimension. The viewport on paper space will allow you to scale, center, rotate, and showcase the drawing located in model space. Later in this chapter, we will explain how to use the viewport window and to establish your Layout (paper space) dimensions. Remember to always draft your drawing in model space, and the title block information will always be drawn or typed into paper space. To return to model space, click on the Model tag on the Command Prompt window (figure 1-4).

Setting Up the Drawing

Units

The first thing to do, before starting a new drawing, is to set the units. On the top toolbar, open **Format**, scroll down, and select Units (figure 1-7). Another way to open the Units pop-up window is to type the word units after the word Command inside the Command Prompt window located at the bottom of your screen.

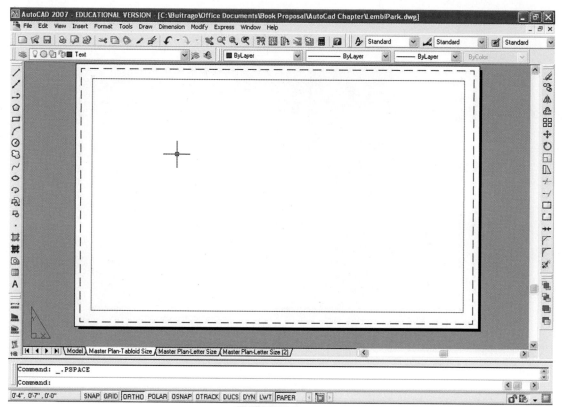

FIGURE 1–6 Paper space screen showing the **Viewports** window frame.

A pop-up window called **Drawing Units** will open (figure 1-8). The **Drawing Units** window will allow you to select Length Type and Precision, Angle and Precision options, and Drag and Drop Scales. Length Type refers to architecture, engineer, and fractional scales of the English measurement system (inches/feet/yards). Decimal and scientific length pertain to the international metric system (millimeters/centimeters/meters). The Angle type units options are decimal degrees, degrees–minutes–seconds, gradients, radians, and surveyor units. For the purpose of this exploration, select Length—Architectural, Precision 0'-00", Angle Type—Decimal Degrees, Precision 0.0, and Drag-Drop Scale—Inches. This will allow you to draw anything in inches either on model or paper space.

---

author's note

It is very important for the first-time CAD users to understand the differences between Architectural and Engineering Length Type. Architectural Length Type uses the unit ratio of 1 inch divided into 16 equal units. Engineering Length Type uses the unit ratio of 1 inch divided into 10 equal units. Architectural Length Type works in inches directly, and Engineering Length Type converts inches into a decimal system. This information will come in handy when typing specific dimensions for an object drafted in model space.

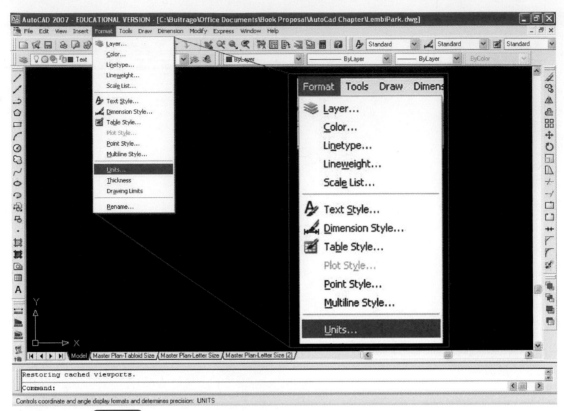

FIGURE 1–7 The **Format** scroll-down menu showing the Units selection.

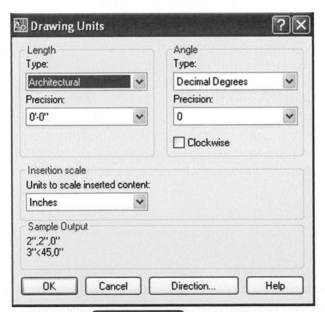

FIGURE 1–8 The **Drawing Units** pop-up menu window.

Limits of drawing—paper size

The next step is to set the limits of your drawing, that is, your paper size. There are several ways to do this. First, click on the Layout tag located on top of the Command

Prompt window (figure 1-4). This will change your screen from model space to paper space. Leave the cursor arrow on top of the Layout tag for a few seconds and right-click on the Layout tag again. This will launch a pop-up window in which you can scroll down and select **Page Setup Manager** (figure 1-9). Once you select the **Page Setup Manager**, another pop-up window called **Page Setup Manager** will open (figures 1-10 and 1-11).

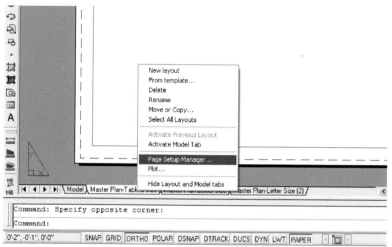

FIGURE 1–9   Pop-up menu window showing the **Page Setup Manager** option.

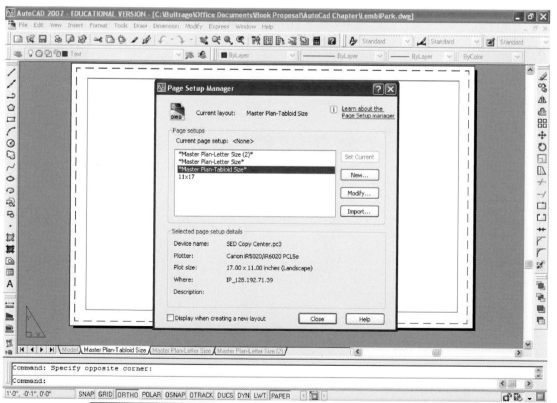

FIGURE 1–10   The **Page Setup Manager** pop-up menu window.

FIGURE 1–11 The **Page Setup Manager** options window.

On the **Page Setup Manager** you will have three options to change the set up of paper space (the size of your sheet). You can select **New**, modify the existing Layout 1 (highlighted in blue), or import another layout from another drawing. For the purpose of this exercise, select **Modify**. A new pop-up window labeled **Page Setup**—Layout 1 will appear (figures 1-12 and 1-13).

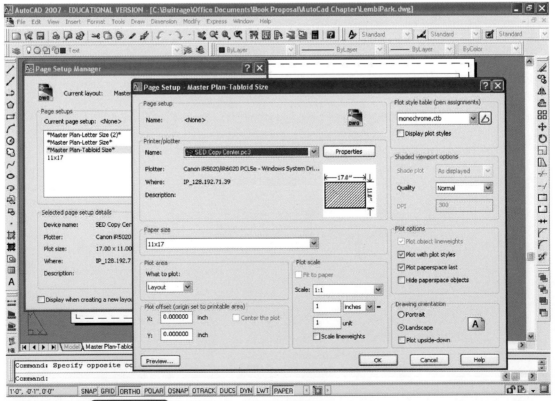

FIGURE 1–12 The **Page Setup** pop-up menu option window.

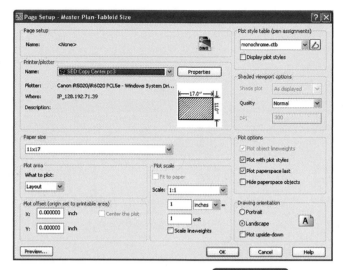

FIGURE 1–13   Setting the options on the Page Setup menu window.

---

author's note

To understand the chapter figures, keep in mind that in the author's CAD drawing the Layout 1, 2, and 3 tags have been renamed as Master Plan—Tabloid Size, Master Plan—Letter Size, and Master Plan—Letter Size(2).

The paper size selection for your drawing will depend on your system configuration for printer(s) selection(s). CAD will automatically list your printer options in the Printer/Plotter—Name window selection option. The default CAD printer option is None, but if you click on the arrow next to this, a pop-up window will list your printer options. Keep in mind that the size of your output (paper sheet) is limited by your system printer(s) selection(s). To create a 24″ × 36″ sheet, you must select a printer that supports this size. If you select a printer that does not support the size of your drawing, a printing/plotting error will occur. The next step is to match the properties of your output with the properties of your printer selection.

After you select your printer, click on the Properties button. This will open another pop-up window called Plotter Configuration Editor (figures 1-14 and 1-15). This window will list your printer Device and Document Setting. Select from that list Modify Standard Paper Sizes (Printable Area), and a window listing all the paper size options for your printer will appear below. Select from that list the size of your final output, and hit the Modify button to the right. Please make a note of how your paper size selection is named in this list. You need to know this in order to match your Paper Size selection in the Page Setup —Layout window, with your printable area. Another pop-up window will open called Custom Paper Size—Printable Area. This window will allow you to set up the printable area (margin limits) of your output (figure 1-16). Make sure that your printer/plotter configuration supports your left, right, top, and bottom margin limits configuration.

As a rule of thumb, most printers will allow you to set up a printable area up to ¼″ or 0.25″ from the edge of the paper. Once your margin selection is made, click on Next, and confirm changes by selecting Finish on the next pop-up window. This will take you back to the **Plotter Configuration Editor** window (figure 1-14).

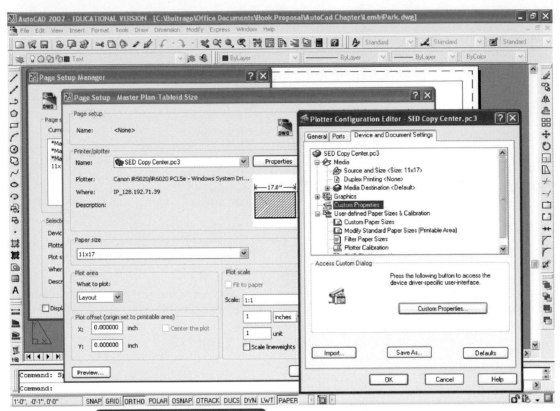

FIGURE 1–14   The **Plotter Configuration Editor** pop-up menu window.

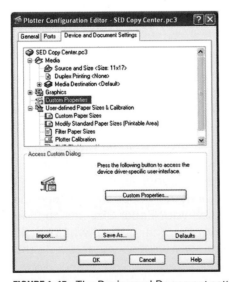

FIGURE 1–15   The Device and Document settings tag.

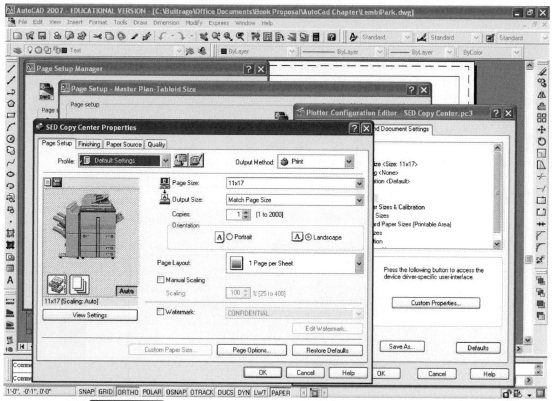

FIGURE 1–16   The Page Setup tag option under the local plotter window settings.

If you are planning to plot your drawing directly to a printer, then hit the OK button. A pop-up window will appear to confirm the changes to your printer configuration file by creating a setup a Plot file document. This is an automatic task of CAD to ensure that your Plot configuration is saved. All you need to do is to hit the OK button. If you are planning to print your document in a Printer Shop such as Kinko's, then you need to create a Plot file document.

To Plot-print a document, select the Ports tag in the Plotter Configuration Editor window. This will open another list where you need to select the Plot File option, and then hit the OK button. Again you will see a pop-up window to confirm the changes to your printer configuration file by creating a setup a Plot file document. This is an automatic task of CAD to ensure that your Plot configuration is saved. All you need to do is to hit the OK button.

Once you are done with this step, CAD will bring you back to the Page Setup — Layout 1 window (figure 1-12). There you need to match your printable area selection with your paper size. On the Paper Size option menu, make sure that the name of your Printable Area selection matches your paper sizes, as previously explained.

The next step is to select Layout from the Plot Area pull-down menu. This will allow you to plot your drawing from paper space. Also, you need to select Scale 1:1 ratio from the Plot Scale pull-down menu. Since the actual scale of the drawing is set up on the viewport; the actual scale of the plot is 1:1. We will explain how to set up the Viewports scale later in this chapter.

In the **Page Setup** —Layout 1 window, another selection option is called the Plot Style Table (Pen Assignments). Selecting this option will open a pull-down menu where you can select the "lineweights" of your plot. Keep in mind that lineweights pertain to the saturation or amount of black ink. The higher the saturation of black ink, then the heavier the lines appear. This pull-down menu will also allow you to select other options like your own Pen Assignment, ACAD, Monochrome, Greyscale, Screening Percentages, and others (figure 1-17).

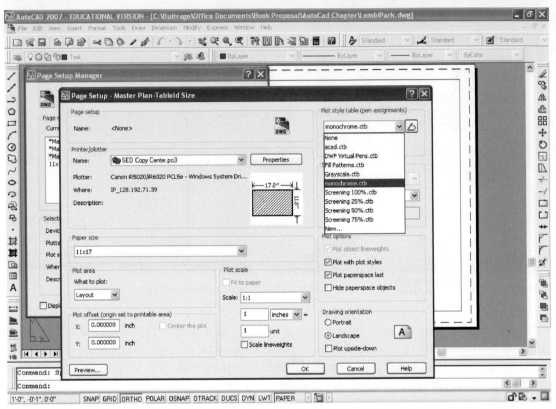

FIGURE 1–17 Setting the Plot Style Table (Pen Assignments).

The Shade viewport option will allow you to select the output quality of drawing. Most drawings will be plotted in the Normal option, but if you need to save ink, draft mode works well. Maximum is the best quality output but will take more time to print (figure 1-18).

Drawing Orientation must match your Paper Size properties. Depending on the size and printer configuration, choose between Landscape or Portrait view. To prevent printing errors, select Preview to see your finished output configuration. After this, to return to the **Page Setup** —Layout 1 window, hit ESC. Once you are satisfied with your **Page Setup** —Layout options, hit the OK button. This will bring you back to the Page Manager window (figure 1-11). Then hit the **Close** button to complete the Page Setup selection process. Now, your paper space tag window will show your page setup settings.

If you need to create another page setup (different paper size), just right-click on the next tag (Layout 2) and repeat the process explained above. If you need to create more Layout Tags, just right-click on the tag to get the pull-down menu option and select from New Layout, Move or Copy, Rename the Tag, and others. See the preceding Author's Note.

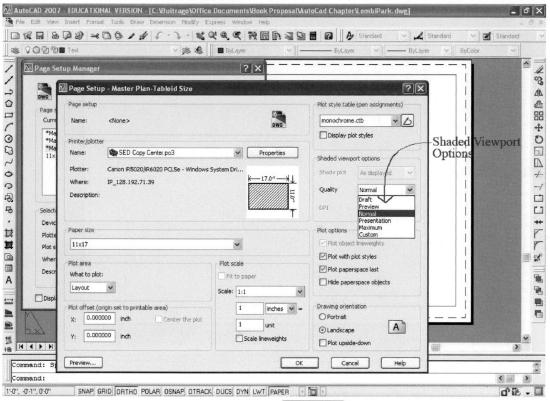

FIGURE 1–18 Setting the print quality under the shaded **Viewports** options scroll menu.

CAD also offers you the option to select the **Page Setup Manager** option from the top toolbar of your screen. Just look under **File**, scroll down the menu to the **Page Setup Manager** option, and follow through. Also, Plot Preview is found under **File** as well (figure 1-19).

author's note

A note of caution: depending on the default settings of your computer, you need to make sure that the default setting of your printer matches those of your CAD **Page Setup**. You can check this by looking under the **Printers and Faxes** option under the Start button on your computer screen (not the CAD screen). See figure 1-20. Please make sure that the correct printer is selected as the default printer, and that the properties of that printer (page size and paper orientation) match those on the CAD **Page Setup** options. This will prevent plotting a CAD drawing to a printer that does not support the size or orientation of your plot. Setting all the details of the plotting configuration at this stage will ensure that the final output will print and appear exactly as seen on the screen. A test plot is seldom recommended (per CAD user discretion) at this stage to troubleshoot any further modifications to the drawing.

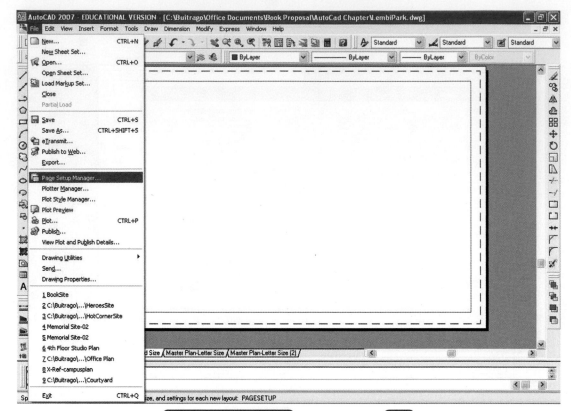

FIGURE 1–19 Selection of the `Page Setup Manager` option under the `File` tag at the top menu bar.

Layers

The Layer Properties Manager

Return to Model space by clicking on the Model tag on the Command Prompt window (figure 1-4). On the upper right corner of the top toolbar, you will see a tool icon that looks like a stack of paper. This is called the `Layer Properties Manager`. Click on this icon and a pop-up window called the `Layer Properties Manager` will open (figures 1-21 and 1-22).

The CAD `Layer Properties Manager` will allow you to create and organize your drawing more efficiently by assigning properties such as lineweights, line thickness, line-types, images, hatch patterns, text, color, and many others by grouping them in specific layers. This will also facilitate further manipulation of your drawing by isolating a specific layer and changing its properties. Layers are organized on a table format. On the horizontal matrix they are classified by Status, Name, On, Freeze, Lock, Color, `Linetype`, `Lineweight`, Plot Style, `Plot`, Current Viewport Freeze, New Viewport Freeze, and Description. On the vertical matrix of this table, layers are listed by name. On the top of the Status column, you will see three icons labeled New Layer (yellow star and paper), Delete Layer (red cross mark), and `Set Current Layer` (green checkmark). The default first layer is Layer 0. To create a new layer, click on the New Layer icon, and the new layer by the default name Layer 1 will be created.

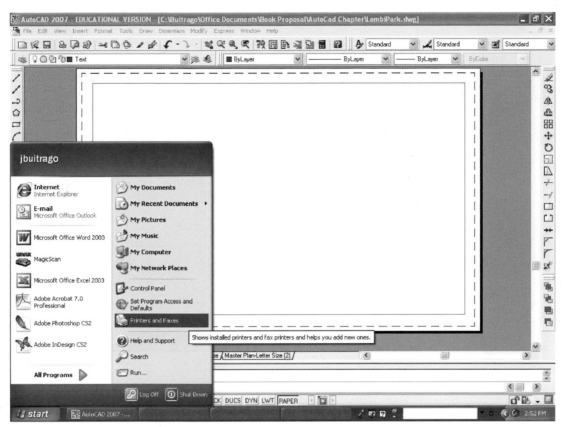

FIGURE 1–20   The Printer and Faxes tag under the Start button of the windows toolbar.

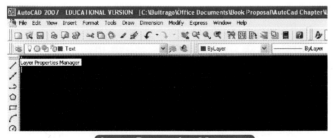

FIGURE 1–21   The **Layer Properties Manager** icon/tool location.

author's note

It is recommended the name of the layers be changed to accurately describe what is on each layer that you create. This will help you greatly to keep your drawing better organized since there is no limit to how many layers can be created. To change the default layer name, for instance Layer 1, double-click the left button of your mouse on the default layer name. When it is highlighted in blue, type in the new name. Figure 1-22 shows the names and properties of layers created and renamed by the authors (not to be confused with Automatic Default CAD names).

FIGURE 1–22 The **Layer Properties Manager** pop-up menu window.

Changing the Layer Linetype

To modify the properties of each layer, first make that layer the active layer by selecting it and clicking on the **Set Current Layer** icon. A green checkmark will appear on the Status column confirming the action. To change each property of each layer, just click on the columns icon. For example, to change the linetype of the layer, just click on the **Linetype** column of the layer. A pop-up window called **Select Linetype** will open. A list of several linetypes will appear. If only one linetype appears on the list, then you must load more linetypes by clicking on the Load button. Another pop-up window listing several linetypes will appear. Click on the one you need, and then hit the OK button. This action will return you to the **Select Linetype** window, which now shows, on the scroll-down list, the linetype that you just loaded. To complete the action, just click on the linetype, and hit the OK button. Your linetype now is permanently assigned to that layer in the Layer Properties (figure 1-23).

Changing the Layer Lineweight—Thickness

To change the layer lineweight, just click on the **Lineweight** column of the layer you want to change. A pop-up window called **Lineweight** will open. Scroll down to your desired lineweight, select it, and then hit OK (figure 1-24). The CAD user can specify different line thicknesses with the same color when plotted in the Monochrome Pen Assignment (figure 1-25).

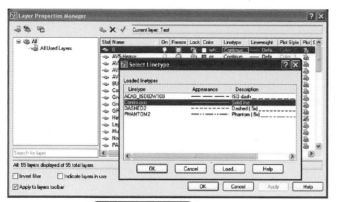

FIGURE 1–23 The `Select Linetype` pop-up menu window options.

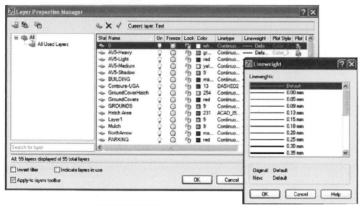

FIGURE 1–24 The `Lineweight` pop-up scroll menu option window.

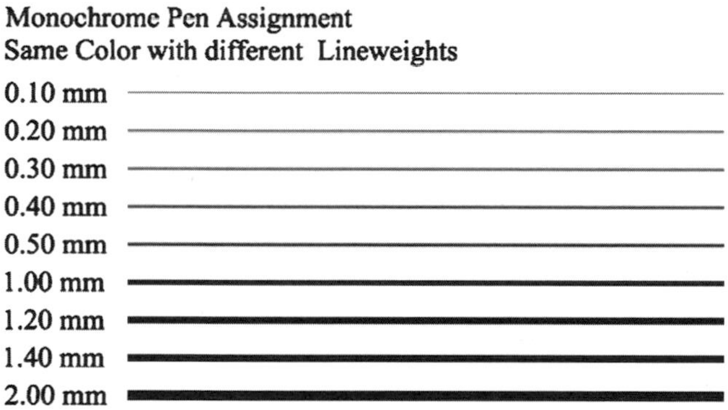

Monochrome Pen Assignment
Same Color with different Lineweights

0.10 mm ——————————————————————

0.20 mm ——————————————————————

0.30 mm ——————————————————————

0.40 mm ——————————————————————

0.50 mm ——————————————————————

1.00 mm ——————————————————————

1.20 mm ——————————————————————

1.40 mm ——————————————————————

2.00 mm ——————————————————————

FIGURE 1–25 The Monochrome Pen Assignment.

Changing the Layer Color

Color in CAD is used to visually organize your drawing. For example, if you create a layer and label it 10 Feet Topo Lines, and make it red, then you can assume that anything red in your drawing will be a 10 Feet Topo Line. Also, color might be used to determine the lightness or darkness of the line. When you click on the Color column of the Layer Properties, the **Select Color** window will open (figure 1-26). You can choose the color of the layer by selecting it from the AutoCAD Color Index (ACI), which runs from Color Index 10 through 249 (Full Color Palette). Under the Color Index (ACI) table is the Standard Colors: Red 1, Yellow 2, Green 3, Cyan 4, Blue 5, Magenta 6, White 7, Dark Grey 8, and Light Grey 9. The third color selection is the Grey shade gradient, which runs from Color Index 250 Dark Greys through 255 Light Greys (figure 1-27).

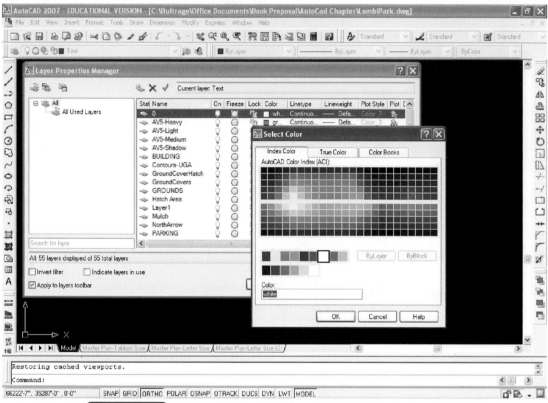

FIGURE 1–26 The **Select Color** pop-up menu option window.

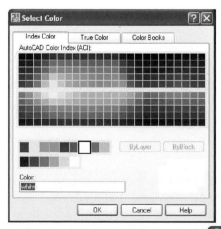

FIGURE 1–27 The Index Color tag on the **Select Color** window.

As previously mentioned, color may be used to determine the lightness or darkness of the line on the final plotted document. To choose this option you must select the Greyscale Pen Assignments from the `Page Setup Manager` window (figure 1-17). In the Greyscale Pen Assignment, Yellow is the lightest, followed by Grey #9, Cyan, Green, Grey #8, Magenta, Red, Blue, and White as the darkest (figure 1-28). This Greyscale Pen Assignment format is used to produce a base drawing that allows landscape architects to draw on it without concealing all the background information. The Monochrome Pen Assignment will give the same lineweight value to all the lines regardless of the layer color assignment (figure 1-29). The Monochrome Pen Assignment is commonly used in landscape architecture practice to print a quick base drawing that was provided by a third party that is not using the office's customized pen assignments. The AutoCAD pen assignment will plot the lines in their original layer color if plot in a color plotter. Otherwise, if a strictly black-and-white plotter is used, it will print in greyscale. The Screening Pen Assignment will plot all the lines and colors on a greyscale value that correlates to the percentage selected. The Screening Pen assignments follow the same landscape architecture drafting purpose of the Greyscale Pen Assignment.

Grayscale Pen Assignment

Yellow
Grey 9
Cyan
Green
Grey 8
Magenta
Red
Blue
White

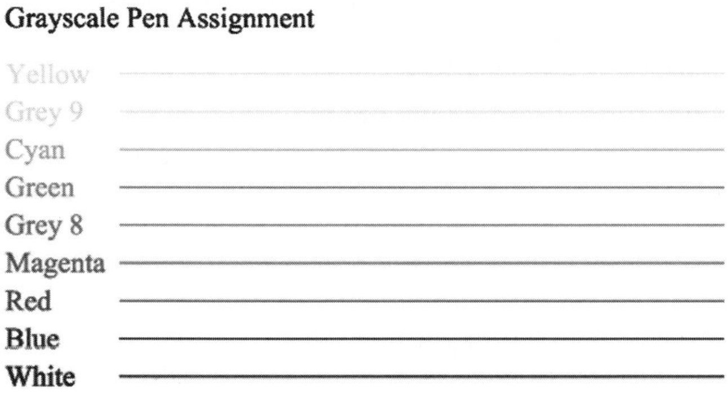

FIGURE 1–28 Greyscale Pen Assignment.

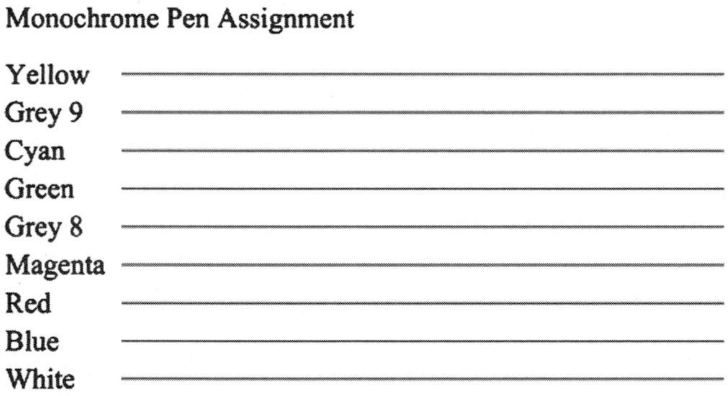

Monochrome Pen Assignment

Yellow
Grey 9
Cyan
Green
Grey 8
Magenta
Red
Blue
White

FIGURE 1–29 The Monochrome Pen Assignment.

The CAD user can create a customized look by combining different lineweights (thickness) with colors (Greyscale Pen Assignment).

The ACI color tables (from 10 through 249) are lighter tones than those from 1 through 9. The grey gradient is the lighter tones from the Color Index. See figure 1-23. In the `Select Color` window, there are two tags labeled True Color and Color Books. Owing to the introductory nature of this book, these two options will be covered in a future advanced-level edition of this textbook.

Layers On, Freeze, Lock, Plot Style, and Plot

The Layer Properties Manager will also allow you to further manipulate your layers. You can turn a layer On or Off by selecting the icon that looks like a yellow light bulb. This action will either display the layer on the screen or not. This will allow the user to work more efficiently by decreasing the confusion of too many objects displayed on the screen, and isolating specific layers to work on. This action will not erase the layer from the drawing and so it will take memory space. That means that your drawing regeneration will be slow (figure 1-30).

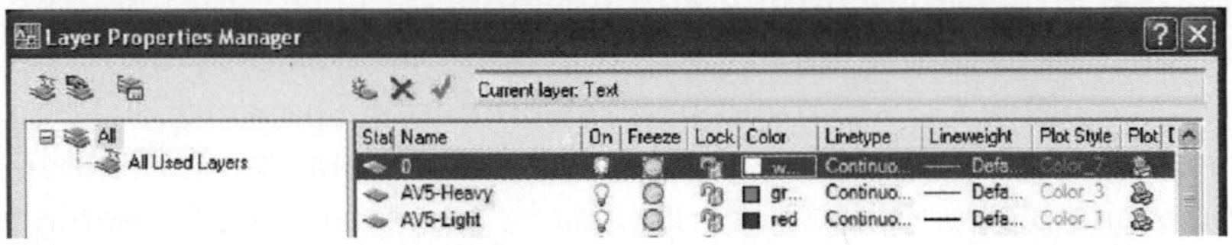

FIGURE 1–30 Turning On/Off layers.

Freeze and Thaw is the icon that looks like a yellow snowflake. The Freeze Layer Tool is similar to turning the layer on or off, except that the memory space that the layer takes from the drawing will not be used to calculate regeneration, thus speeding up this process. Although layers that are turned off and Freeze are not displayed on the screen, they will be plotted.

The yellow Lock icon is used to lock or unlock a layer. This action will prevent the layer from being edited or modified. The layer will still be visible, but you cannot draw in this layer or make changes to this layer. This tool is generally used to prevent accidental modifications to a layer (figure 1-31).

FIGURE 1–31 Locking a layer.

Plot style is simply the pen assignment as described by your color index selection.

The icon that looks like a printer is called Plot . This tool will allow you to plot an entire layer. This makes the manipulation of the final plotted document more efficient by directly selecting which layer should be included on the finished output.

Opening and closing the Properties Layer Manager

Once you create the layers and assign the proper properties to each one, you can hit the OK button. To open the Properties Layer Manager, just click on the icon (figure 1-21). Keep in mind to always work on layers, since this will facilitate further revisions to the properties of your objects. Remember that anything you draw will be located in the current layer.

Properties Layer Managers Status Display windows

Next to the Properties Layer Manager icon (stack of papers) is the Status Display window (figures 1-32 and 1-33). This window will display the current active layer properties such as Layer On, Freeze, Thaw, Lock, Color, and Name. If you want to change to another layer without opening the entire Properties Layer Manager, click on the arrow of the Status Display window, scroll, and select another layer.

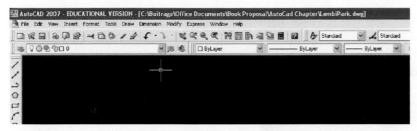

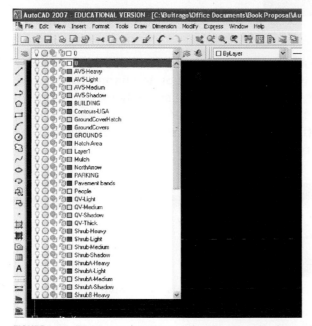

FIGURE 1–32 The Status Display window.

FIGURE 1–33 The Layer (scroll menu) options on the Status Display window.

Next to this window is a tool icon that looks like a stack of paper on top of a yellow sheet. This is called Make Object's Layer Current (figure 1-34). If you want to change an object on your screen to another layer, thus changing its properties as well, click on this icon and click/select the object on the screen. This will change the object to the current layer.

FIGURE 1–34 Make Object's Layer Current.

Next to the Make Object's Layer Current tool is the Layer Previous tool. This tool will allow you to turn back to a previous layer that was set up as current (figure 1-35).

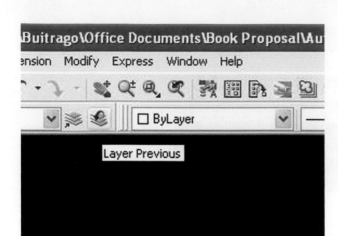

FIGURE 1–35 The Layer Previous tool icon.

Next to the Layer Previous tool icon are Color Control, Linetype Control, and Lineweight Control display windows. These windows will allow you to change the properties of an object regardless of the default layer properties. We strongly recommend not using this format, and keeping them on their default by Layer status. It is best to keep the objects' properties by layer since the layer properties will always override the display windows property at the time of plotting (figure 1-36).

FIGURE 1–36 The Color Control, Linetype Control, and Lineweight Control display windows.

The CAD and the Mouse

When we draw, our eyes follow the motion of the hand while it glides on the surface of the paper. This visual and motion connection is direct, and in simple terms we see what we draw. When we draft on the computer, our eyes are locked on the screen while the hand is holding the mouse, thus breaking the direct visual–motion connection of the drafting process. In simple terms, we do not see what we draw since the eye has lost its direct eye (visual) and hand (motion) connection.

The re-training of the hand-and-eye coordination is something that some beginner CAD users find difficult to master but, with practice, is easily done. To make things a little more complicated, there are now several brands, styles, colors, and types of mouse available in today's computer market such as the laser mouse, optical mouse, and traditional roll ball mouse. The variety of products might be confusing, but in general, any good CAD-compatible mouse must have a middle button or wheel, a left button, and a right button (figure 1-37).

Depending on the CAD tool, clicking on the left button is for selection or engaging. The right button click is for enter or cancel. The wheel will allow the user to zoom in or

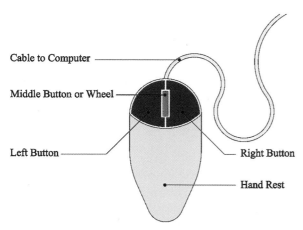

FIGURE 1–37   The standard mouse.

zoom out, and double-clicking on the wheel/button will "zoom" the entire drawing to fit the "extent" of the screen. The hand rest is used to move the mouse and hence the cursor on the screen. Further notes on which button (left, middle, or right) to click or press to operate a particular tool will be explained in the following section.

Standard CAD Tools

The standard CAD tools are the most frequently used tools to draft a drawing. These tools will show on the CAD screen whenever CAD is launched. These tools are usually "nested" in the left or right side bar, and/or in the top bar. They can also be customized per the operator needs. Figures 1-38 and 1-40 show the standard tools icon's location and names. Figure 1-41 shows the standard, default title-top toolbar. To select and/or engage any of these tools, left-click on the icon. To disengage any tool, hit the Escape (ESC) or Enter key. Also, many of these tools have further options that are listed in the Command Prompt window. These options appear in parenthesis () or brackets [] once the tool is engaged.

FIGURE 1–38   The standard tools icon's location.

author's note

The following tool's definitions and explanations will help the CAD user to become familiar with the basic drafting tools. It is recommended that you practice on the CAD screen as you read this section of the chapter.

The *Line tool* will allow the user to draw a line in any length, angle, or direction. To select this tool, left-click on the icon. Then place or pan the cursor on the first point of the line, click with the left button of the mouse to set the point of origin of the line, then drag-stretch the line to the second endpoint location, and left-click to set in place. To draw a line to a specific dimension or length, after the first click, type the number for the length of the line segment in the Command Prompt window, followed by Enter. You can continue drawing a second line that is connected in segments (individual) to the first line by pressing the left button of the mouse to set a third, fourth, and any other points within the line. To disengage the tool, hit ESC or Enter.

Construction Line is a line that goes to infinity in any angle or direction. When engaging this tool, the location of the cursor will set the center of this line.

The *Polyline tool* will allow you to draw a line in segments that are continuously connected. Also, it will allow you to draw curves and straight segments. To shift from straight or curve Polyline, just type Arc or Line in the Command Prompt window.

The *Polygon tool* will create geometric forms of equal-length sides such as triangles, squares, pentagons, hexagons, and octagons. When engaging this tool, pay close attention to the Command Prompt window, since it will require specifying (type) the numbers of sides (from three to infinity), selecting the center point or edge length, and inscribing the location.

The *Rectangle tool* will allow you to draw a rectangle of any area, dimension, and rotation. Select your preference by typing the command on the Command Prompt window.

The *Arc tool* works in the same way as the line except that after setting the first point-beginning of the line, the second point will establish the center of the arc, followed by the third endpoint of the arc line.

The *Circle tool* will allow you to draw a circle of any area, center point, radius, diameter, and tangent point. Select your preference by typing the command on the Command Prompt window.

The *Revision Cloud* will draw a multibubble enclosed line. Select your preference for arc length, object, and style by typing the command on the Command Prompt window.

The *Spline tool* will let you draw a freehand soft curve. The disadvantage of using the Spline tool is that the shape created with this tool will not allow the use of the hatch, area, and dimension tools.

The *Ellipse tool* will let you draw an ellipse shape of any area, center point, radius, and diameter. Select your preference by typing the command on the Command Prompt window.

The *Ellipse Arc tool* will let you to draw an open, not enclosed, ellipse shape of any area, center point, radius, and diameter. Select your preference by typing the command on the Command Prompt window.

The *Insert, Make a Block, Point, Hatch,* and *Gradient tools* will be covered later in this chapter.

The *Table tool* will allow you to create a Matrix of rows and columns of your preference. Just follow the pop-up menu windows and select your preferences accordingly.

The *Multiline Text tool* will let you type and create text. When engaging this tool, a pop-up menu window (Text Formatting) will open which looks like any word document window. This will let you select font type, style, size, color, and paragraph justification. Select your options according to your preferences (figure 1-39).

FIGURE 1–39 The Text Formatting pop-up menu window.

The standard right-side toolbar of the authors screen (figure 1-40) shows the Erase, Copy, Mirror, Offset, Array, Move, Rotate, Scale, Stretch, Trim , Extend, Break at Point,

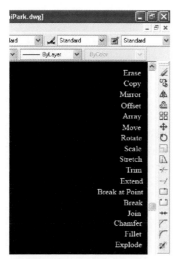

FIGURE 1–40 The standard tools icon's names.

Break, Join, Chamfer, Fillet, and Explode tools. To engage any of these tools, left-click on the icon. To disengage any tool, hit ESC or Enter. Also, many of these tools have further options that are listed in the Command Prompt window. These options appear in parenthesis () or brackets [] once the tool is engaged. These tools are as follows:

The *Erase tool* will allow you to erase any object from the screen. After engaging this tool, left-click on the object and hit Enter, or right-click on the mouse. The right clicks on the mouse and the space bar on your keyboard work the same as Enter.

The *Copy tool* will let you copy objects as many times as needed. Engage the tool with a left click on the mouse. Then left-click on the object to be copied and hit Enter (or right-click the mouse) to select it. The Command Prompt window will show the Base Point and/or Displacement option. Pan the cursor to the location of the base point (displacement reference point) and then left-click to set. Now the copied object will show as a dashed line. Using the cursor, pan around for placing the object in the screen location of your preference and left-click to set in place. Another copy of the object will appear; left-click to set in place. To disengage the tool, just hit ESC or Enter.

The *Mirror tool* will allow you to create a mirror image of the object selected along an axis. Engage the tool with a left click on the mouse. Then left-click on the object to be mirrored and hit Enter (or right-click with the mouse) to select it. The Command Prompt window will read "Specify first point of mirror line." Pan the cursor to the location of the first point (first point of axis line) and then left-click to set. Now the Command Prompt window will display "Specify second point of mirror line." Pan the cursor to the location of the second point (second point of axis line) and then left-click to set. The Command Prompt window will show "Erase source object? [Yes/No]." Type Yes or No as per your preference and hit Enter (or right-click with the mouse). A reverse copy (mirror) of your selected object has been created along a reference axial line.

The *Offset tool* will let you make copies of the selected object at a specified offset distance from the selected source. Engage this tool by left-clicking on the tool icon. The Command Prompt window will show "Specify offset distance." Type the distance per your preference and hit Enter. After this action, the Command Prompt window should say "Select object to offset." Left-click on the object to be offset. The Command Prompt window now will read "Specify point on side to offset." Left-click (on the screen) outside or inside the object to specify the side in which the object will be copied. An exact duplicate of the selected object will appear at the specific offset distance and with the same point of reference (origin) from the source.

The *Array tool* will allow the CAD user to create multiple copies of the same object arranged in a rectilinear (Rectangular Array) or concentric (Polar Array) matrix of any number of rows, columns, angles, and number of copies. This tool will be fully explained later in this chapter.

The *Move tool* will let you select an object (left click and right click) and move it to any location in the screen.

The *Rotate tool* will allow the user to rotate any object. To engage this tool, left-click on the tool icon. Then select the object to rotate (left click and right click).

The Command Prompt window now reads "Specify base point." Left-click on the screen to set the base-origin point (pivot) of rotation. The Command Prompt window will now read "Specify rotation angle or [Copy/Reference]." Type the angle number or pan with the cursor to rotate the object to the desired angle. Left-click to set in place.

The *Scale tool* will allow the user to scale up or down an object. Engage this tool by left-clicking on the tool icon. Then select the object (left and right click). The Command Prompt window now reads "Specify base point." Left-click on the screen to set the point of origin/reference where the selected object will be enlarged or reduced. The Command Prompt window now reads "Specify scale factor or [Copy/Reference]." Type the scale factor of your preference. To make the object twice as big, type number 2, type 3 for three times larger, and so forth. To make it smaller, type 0.5 for half the size, 0.25 for one quarter smaller, and so forth.

The *Stretch tool* is used to literally stretch-elongate an object. To engage the tool, left-click on the tool icon. On the screen, left-click and hold the button down. Drag the cursor to create a dashed, green filled polygon and stretch to cover the sides of the object to be stretched and right-click to select. The Command Prompt window now reads "Specify base point or [Displacement]." Left-click on the screen to set the reference point and left-click again. Move the cursor away from the object in the desired direction to stretch the object and left-click to set in place.

The *Trim tool* is used to cut objects or lines by using an intersecting/overlapping reference object or line. Left-click to engage this tool. The Command Prompt window now reads "Select object." Left- and right-click on the reference object to be used as "the cutting edge." Left-click on the objects to be cut or trimmed at the reference object/line. Hit ESC or Enter to disengage this tool.

The *Extend tool* will let you stretch a line to a reference object or line. Left-click on the icon to engage this tool. Left- and right-click to select the reference point where the lines are to be stretched. Left-click on the lines to extend them to meet the reference object or line. Hit ESC or Enter to disengage this tool.

The *Break at Point tool* will allow the CAD user to break a continuous line, an enclosed polyline, and/or an enclosed object of any shape and form at a specific point. To understand how to use this tool, just draw a simple rectangle on the screen. Left-click on the tool icon to engage. Select object to break by left and right clicks. Then left-click on the desired location were the enclosed polyline object will break into two.

The *Break* works in the same fashion as the Break at Point, except that it will create a gap between the endpoints of the break location.

The *Join tool* will let the user join together supported objects.

The *Chamfer tool* is used to create a cut between two lines at a 45° angle and at equal distance from the original intersection point of the two lines. To engage this tool, left-click on the tool icon. The Command Prompt window now reads "Select first line or [Undo/Polyline/Distance/Angle/Trim/method/Multiple]." Type D for distance. Next specify the distance for the first line from the intersection point and hit Enter. Type the second point distance and hit enter. The Command

Prompt window now reads "Select first line." Left-click on the first line, followed by a second left click on the second line to chamfer. A 45° segment line connecting the original intersecting line will be created.

The *Fillet tool,* like the Chamfer tool, is used to round off the intersection point between two lines with an arc-curve. To engage this tool, left-click on the tool icon. The Command Prompt window now reads "Select first line or [Undo/Polyline/Radius/Trim/Multiple]." Type radius, hit Enter, and then specify radius distance and hit Enter again. The Command Prompt window now reads again "Select first object or [Undo/Polyline/Radius/Trim/Multiple]." Just left-click on the first line, followed by a second left click on the second intersecting line. An arc with the specified radius connecting both lines will be created.

The *Explode tool* will allow the user to break in segments a continuous polyline, object, group of objects, and/or block. Engage this tool with a left click, followed by the Command Prompt window settings.

The title-top toolbar (figure 1-41) shows the standard, default functions of the CAD document such as `File`, `Edit`, `View`, `Insert`, `Format`, `Tools`, `Draw`, `Dimension`, `Modify`, `Express`, `Window`, and `Help`.

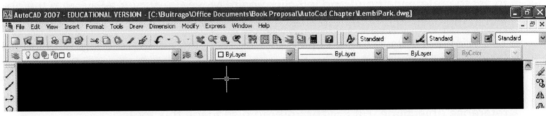

FIGURE 1–41 The title-top toolbar.

Anyone familiar with the operation of a Windows program will recognize `File` and `Edit` as the tools used to manage the document. A left click on the `File` tool icon will open another pop-up menu window with more options (figure 1-42). Several of the basic functions of `File` and `Edit` have been already explained at the beginning of this chapter. `View`, `Insert`, `Format`, `Tools`, `Draw`, `Dimension`, `Modify`, `Express`, `Window`, and `Help` will be further explained within the context of creating your first drawing.

Drafting Settings

The `Drafting Settings` will allow the user further control and manipulation of the drafted objects on the screen. The tabs to enable the drafting settings are located in the status bar under the Command Prompt window (figure 1-43). These settings are Cursor Coordinates, Snap, Grid, Ortho, Polar, Osnap, Otrack, DUCS, DYN, LWT, and Model/Paper.

The Snap tab is used to specify the gap or distance between the line ends drafted on the screen automatically. To set the Snap tool to a specific option, left-click on the tab to turn Snap on, then right-click. This will launch a pop-up menu window, and from there select Settings. This will launch the `Drafting Setting` pop-up menu window for `Snap and Grid` options. Set the Snap spacing on X and Y to 0 to make the lines continuous or specify a separation distance (figure 1-44). After selecting your preferences, hit the OK button.

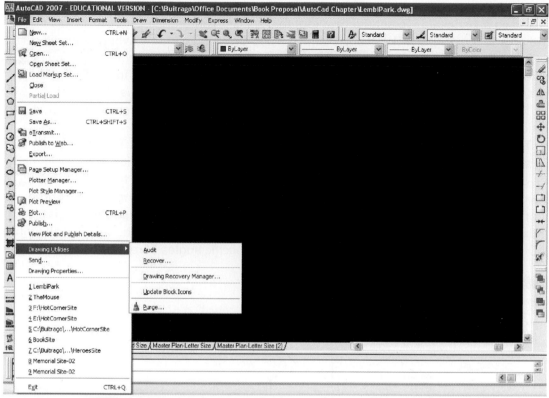

FIGURE 1–42 The File tag.

FIGURE 1–43 The Command Prompt window and tabs options.

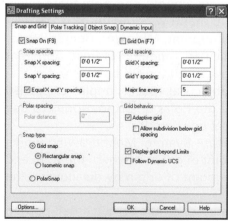

FIGURE 1–44 The Drafting Setting and the Snap and Grid tag options.

The Grid tab will make a nonplot grid (made of points) appear on the screen. This grid is used as a visual aid. To set the Grid tool to a specific option, left-click on the tab to turn Grid on and then right-click. This will launch a pop-up menu window, where you will select Settings. This will launch the **Drafting Setting** pop-up menu window for **Snap and Grid** options. Set the Grid spacing on X and Y your setting preferences (figure 1-44). The default spacing for the grid is ½". After selecting your preferences, hit the OK button.

The Ortho will allow drawing lines and objects on a 90° angle on a horizontal and vertical orientation.

The Polar tab will enable you to draw lines on a radial 360° pattern. To set the Polar tool to a specific option, left-click on the tab to turn Polar on and then right-click. This will launch a pop-up menu window, and from there select Settings. This will launch the **Drafting Setting** pop-up menu window for **Polar Tracking** options. Set the Polar tracking to the desired polar angle—increment angles or **Object Snap** tracking settings to orthogonal or polar (360°) angles (figure 1-45). After selecting your preferences, hit the OK button. This will allow the CAD user to draw lines or objects on the screen at specific angles of reference.

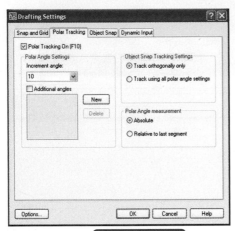

FIGURE 1–45 The **Polar Tracking** tag option.

The Osnap tool will enable you to connect (snap) lines and objects to a specific point of reference. To set the Snap tool to a specific option, left-click on the tab to turn Snap on, then right-click. This will launch a pop-up menu window, and from there select Settings. This will launch the **Drafting Setting** pop-up menu window for **Object Snap** (figure 1-46). The object snap options or modes are Endpoint, Midpoint, Center, Node, Quadrant, Intersection, Extension, Insertion, Perpendicular, Tangent, Nearest, Apparent Intersection, and Parallel. After selecting your preferences, hit the OK button.

The Otrack tab will enable the CAD user to draw objects at specific angles or precise positions using a temporary alignment path such as the Polar or Ortho mode. To use this feature, the Grid, Polar, or Ortho mode must be set on.

DUCS stands for Allow/Disallow the User Coordinate System. Most drawing works on the World Coordinate System (WCS), where the point of origin is where the X, Y, and Z axes intersect at zero. The CAD user can modify the point of origin by disallowing the WCS and setting a User Coordinate System (UCS).

Setting on the DYN or dynamic input will make appear on the screen small pop-up menu windows that contain data (angle, length, coordinates) pertaining to the object or line drafted on the screen as is drawn or set in place (figure 1-47).

FIGURE 1–46 The **Object Snap** tag option.

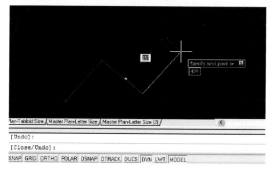

FIGURE 1–47 Setting on the DYN or dynamic input will make appear on the screen small pop-up menu windows.

Setting On or Off the LWT (Show/Hidden Lineweight) will enable the CAD user to modify and change the lineweight as previously explained at the beginning of this chapter.

The Model tab will enable the user to shift the drawing from model space to paper space.

---

author's note

CAD will enable the user to turn on or off the **Drafting Settings** by using the Function keys on the keyboard. The Function keys shortcut functions are:

F1 will access the AutoCAD On-Line help.

F2 will open the **Edit** information window.

F3 will turn the Osnap on or off.

F4 will turn the Tablet tool on or off.

F5 will set the Isoplane to Top, left, or right.

F6 will control Dynamics (UCS).

F7 will turn the Grid on or off.

F8 will control the Ortho mode.

(_Continued_)

F9 will turn the Snap mode on or off.

F10 is for setting the Polar mode on or off.

F11 will control the Otrack settings.

F12 will turn the DYN settings on or off.

Properties Palette

The 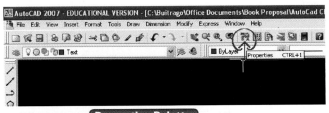 tool icon, on the CAD Top toolbar, is next to the Zoom tool. The **Properties Palette** tool icon looks like a hand with a bar of colors, stack of papers, and the letter A (figure 1-48). Selecting this tool (left-click the mouse) will launch the **Properties Palette** pop-up menu window (figure 1-49). This tool will enable the CAD user to list the current settings for the object or line selected on the screen. It also allows the user to change and directly modify the properties (color, width, linetype, scale, thickness, hatches, angles, settings, etc.) of any object or line selected on the screen. It can be opened and closed any time per CAD-user discretion and can be simultaneously used while drafting an object on the screen.

FIGURE 1–48 The **Properties Palette** tool icon.

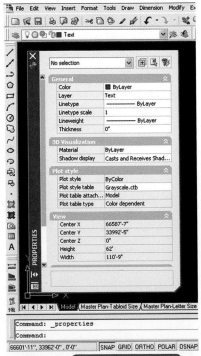

FIGURE 1–49 **Properties Palette** pop-up menu window.

Creating Your First Drawing

The previous segments of this chapter familiarized the readers with all the terms, notions, tools, and settings of CAD. Understanding the basic principles of how to set a drawing before commencing drafting is a necessary first step for successful drafting outcome. It is the authors' research and teaching experience that led to this conclusion of giving first an overall CAD introduction, since it provides the user with a strong technical understanding and foundation. This is also the rule of the land in Landscape Architecture design practice, since every single drawing must follow the adopted office setting format, before drafting commences. So, let us get started.

Start by launching AutoCAD as explained at the beginning of this chapter. By now you should be aware that the default AutoCAD screen open is in model space. Your second action is to set the units of the drawings. Please refer to the "Setting up the drawing" section of this chapter for further details. Third, select your paper size options as described at the beginning of this chapter. For the purpose of illustrating the following exercise, the author's screen will be set up for "architects units" and the paper size to "letter" on a "landscape view" orientation (figures 1-50 and 1-51).

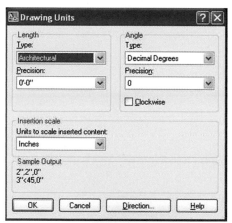

FIGURE 1–50 The Drawing Units pop-up menu window.

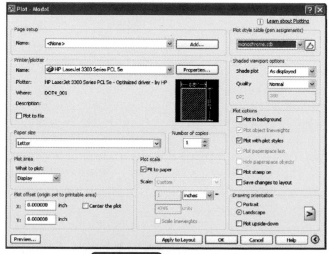

FIGURE 1–51 The Plot Setup pop-up menu window.

Readers are aware that in model space there are no limits in terms of "space." If you draw a line of one inch or one light year . . . it will fit on the screen. Just zoom extent (double-click the wheel of the mouse) to refit the line within the boundaries of the screen. To set the actual scale, a viewport must be created on paper space. Setting the scale of the drawing will be discussed later in this chapter.

The next action to take is to create several layers with different lineweights, color, and thicknesses. The number of layers or the names are determined by the user needs. For the purpose of this exercise, we will create a basic tree-shaped symbol, and per authors' choice, *Quercus virginiana* (live oak) will suffice. To add depth and detail to the tree symbol, several layers were created and labeled in an easily recognizable format (figure 1-52).

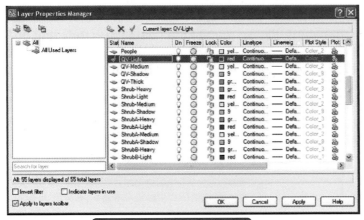

FIGURE 1–52   The **Layer Properties Manager** window.

The author's layers are as follow; QV-Light (Red color), QV-Medium (Yellow), QV-Thick (Green), and QV-Shadow (Grey). Working on layers will let the CAD user manipulate the drawing more efficiently and enhance the graphic quality and depth of the composition by controlling the type, thickness, and color of the line. The next step is to set the QV-Light layer as the active layer and close the **Layer Properties Manager** pop-up menu window. The next question is how big (wide) this tree needs to be. The authors decided that a 22′ wide specimen would do; so start by drawing an 11′ radius circle on the screen as the base of the drawing (figure 1-53). Using the Polyline tool, draw a zigzag pattern line around the circumference of the circle. When the line is close to the beginning point, hit Enter to close the polyline (figure 1-54). The jagged line will give the tree symbol a more interesting and organic feel. Using the Offset tool, make an offset copy of the original zigzag line, which was drawn on the QV-Light layer (red), 6″ apart. Then change the second line to the QV-Thick layer (green) by selecting the line and clicking on the QV-Thick layer (green) on the Layer scroll-down pop-up menu (figure 1-55).

You can further embellish your tree symbol by adding a small circle at the center and/or radial lines. Make sure that the QV-Medium layer (yellow) is the active layer; otherwise, create the lines and circle first, then select the lines to be changed to the QV-Medium layer by clicking on the proper layer at the scroll-down pop-up Layer menu (figure 1-56).

Another way to further embellish your tree symbol is by adding a shadow. Make the QV-Shadow layer the active layer by clicking on this layer on the Layer scroll-down pop-up menu window (figure 1-57). At this point, the original 11′ radius circle is no longer

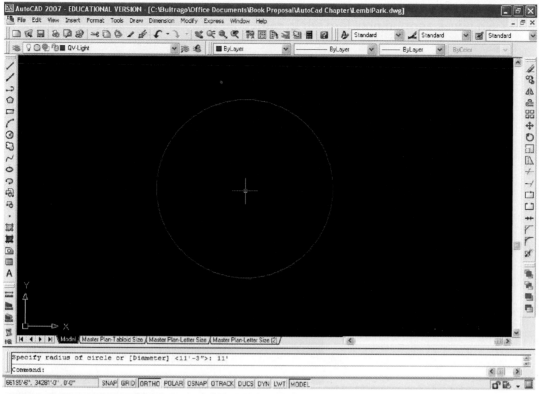

FIGURE 1–53 A 11′ radius circle.

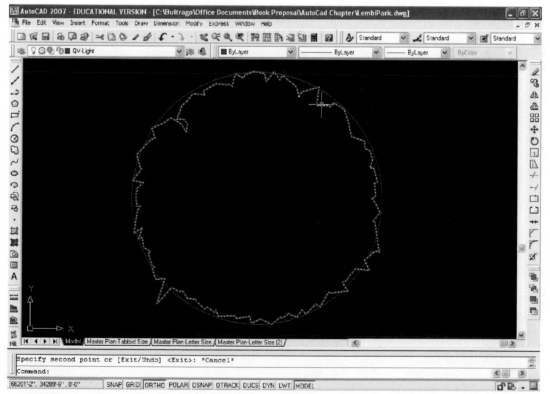

FIGURE 1–54 A zigzag pattern line around the circumference of the circle.

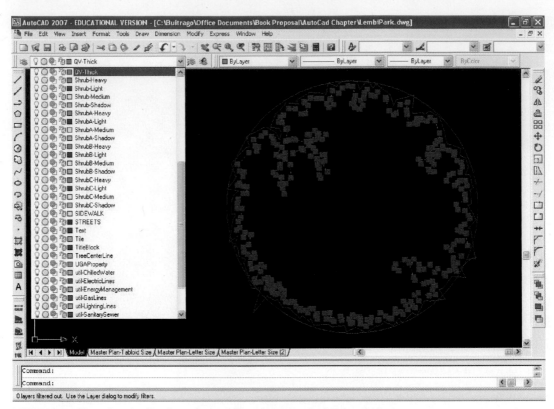

FIGURE 1–55 Changing the line layer by scrolling down on the Layer window.

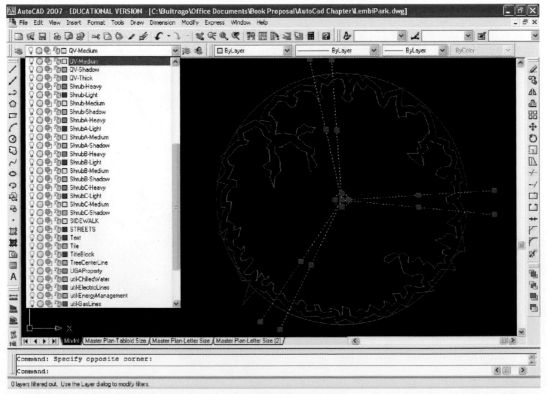

FIGURE 1–56 Changing the radial lines to the QV-Medium layer.

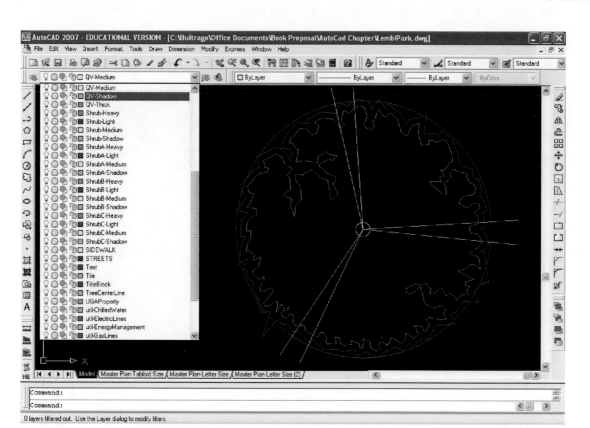

FIGURE 1–57 Select and make QV-Shadow the active layer.

needed, so it can be deleted. Use the erasing tool or just select the line and hit the delete key on your keyboard. Then select the zigzag red line and offset 6″. Next, draw a line across the entire tree symbol, and use the **Trim** tool to cut half of the copied zigzag line (shadow) (figures 1-58 and 1-59).

Delete the trim line and change the half shadow off set line to the QV-Shadow layer. Change the thickness of the half shadow line to create the illusion of a shadow. Just type the command "pedit." The Command Prompt window now will read "Select polyline or [multiple]." Click with the left button of the mouse to select the shadow line, and the Command Prompt window now reads "Enter an option [Close/Join/Width/Edit Vertex/ Fit/Spline/Decurve/Ltype gen/Undo]." Type W for width, and in the next Command Prompt window type 3″. This will change the width of the shadow line, thus creating the feel of a shadow (figure 1-60).

This format will allow the CAD user to create an extensive library of graphic tree symbols that can be copied and used extensively throughout the drawing. CAD will also let the user group these layers into one object without altering the linetype, name, colors, or thickness of the symbol. This tool is called **Block**.

Creating Blocks

Writing a block will allow the CAD user to group several objects into one. There are several ways to access this tool. The **Block** tool can be accessed by typing "wblock" in the Command Prompt window (figure 1-61).

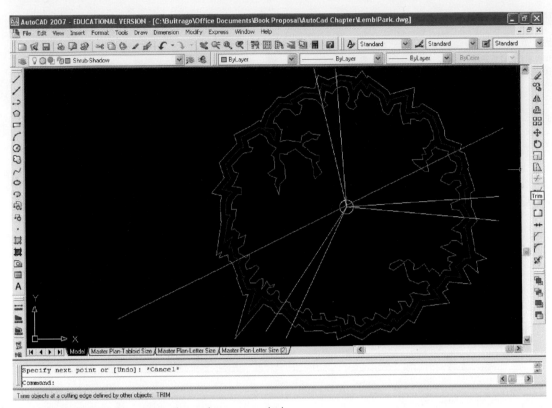

FIGURE 1–58 Draw a line across the entire tree symbol.

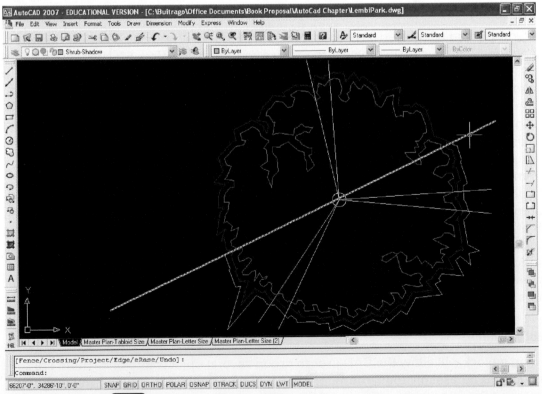

FIGURE 1–59 Use the **Trim** tool to cut half of the shadow off.

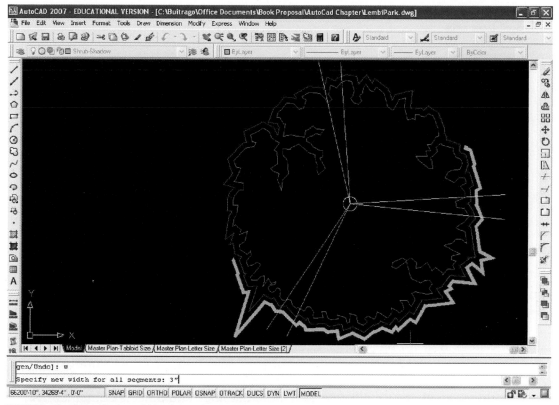

FIGURE 1–60 Change the width of the shadow line.

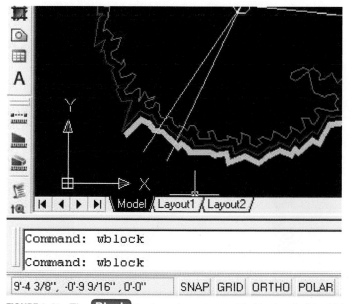

FIGURE 1–61 The Block tool can be accessed by typing "wblock" in the Command Prompt window.

This will launch the Write Block pop-up menu window (figure 1-62). Under the Source option, select objects. Then click the Select Objects icon under the Base option. This will return you to model space.

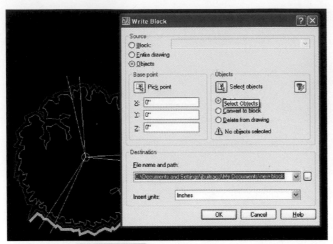

FIGURE 1–62 Write Block pop-up menu window.

On In model space, place the cursor outside the object and click and hold the left button of the mouse. Drag while holding the left button of the mouse to create a dashed-line square with a green field inside. This is called a selection window and everything that is covered by the green field will be selected. Also, notice that the Command Prompt window now reads "Select object: Specify opposite corner" (figure 1-63). Click one more time on the left button of the mouse to make your selection, and then right-click or hit Enter. This will return you to the Write Block pop-up menu window (figure 1-64). Use of the selection tool will ensure the selection of all the lines and objects contained within the boundaries of the square (dash line) area selection. For further details regarding the Selection tool, please refer to the Standard CAD Tools section of this chapter.

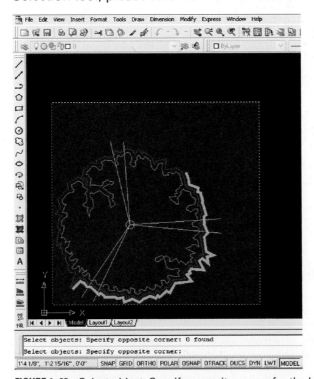

FIGURE 1–63 Select object: Specify opposite corner for the block.

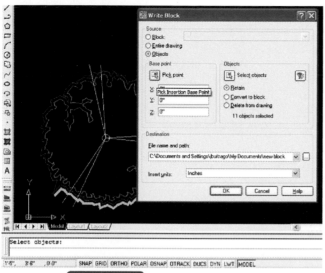

FIGURE 1–64 (Write Block) pop-up menu window.

On the (Write Block) pop-up menu window, under the Base Point option, left-click on Pick Point. This action will return you to model space. Notice that the Command Prompt window now reads "Specify insertion base point." Place the cursor as close as possible to the center of the tree symbol. This will allow the user to establish a common reference point for inserting block since the cursor arrow will always point to the center of the object. Left-click to return to the (Write Block) pop-up menu window. Under the Destination option, look at the filename and path. Click inside this window and give a block a name. The authors chose to call this symbol QV-22 for quick reference to a 22-feet-diameter *Quercus virginiana* (live oak). Next to this window, there is a square button. Click on this to browse and select the folder or file to save your block symbol (figure 1-65) and hit the Save button. Also, under the Units option, select the unit system according to the drawing units system. After completing these steps, hit the OK button. The (Write Block) pop-up menu window will close, and at the upper left corner of

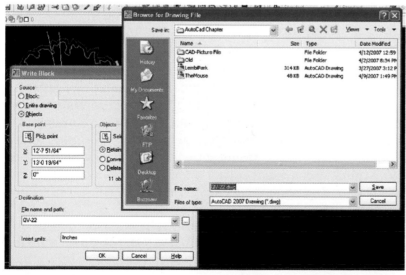

FIGURE 1–65   Click on the square button to browse and select the folder or file to save your block symbol.

your screen, another pop-up window will appear showing a preview of the block created. This preview window appears and closes very fast (no more than 5 seconds).

Another way to access the **Write Block** tool is to look at the top toolbar under **Draw**, scroll down and select **Block**, then select **Make** (figure 1-66). This will launch the **Write Block** pop-up menu.

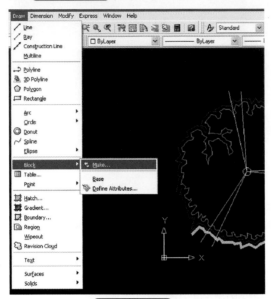

FIGURE 1–66 The **Write Block** tool can be accessed under the top-bar tool tag **Draw**.

After completely creating a block, you may notice that the original objects are still not arranged together as one object. The block that you just created is saved in the file that you previously specified on the **Write Block** pop-up menu window. To access this block, the next action is to insert the block into the drawing.

Insert a Block

There are two ways to insert a block. The first step is just typing the word Insert in the Command Prompt window (figure 1-67). This will launch the **Insert** pop-up menu window (figure 1-68).

On the **Insert** pop-up menu window, hit the Browse button. This will launch your browser pop-up menu window; from there find the file location of the block, select it, then hit the OK button (figure 1-69). This will return you to the **Insert** window. Under the Path option, click (green checkmark) on the Insert Point (Specify On-Screen). That will make the insertion point of the block the same as the insert point (center of the tree symbol). Leave the Scale (Specify On-Screen) and the Rotation unchecked (figure 1-68). Click on the OK button; the **Insert** window will close and the block object will appear. Notice that the cursor is the center of the object and will follow the motion of the mouse. To set the block in place, just click on the left button of the mouse (figure 1-70).

Another way to access the **Insert** tool is to scroll down and select **Block** on the top toolbar under **Insert** (figure 1-71). This will launch the **Insert** pop-up menu.

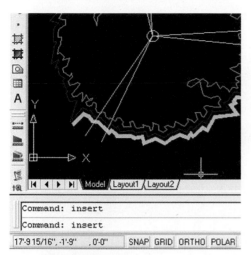

Command: insert
Command: insert

17'-9 15/16", -1'-9"   , 0'-0"   SNAP GRID ORTHO POLAR

FIGURE 1–67 Type the word Insert in the Command Prompt window.

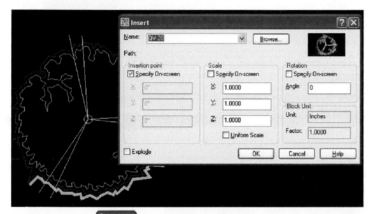

FIGURE 1–68 The Insert pop-up menu window.

FIGURE 1–69 Hit the browser button to launch the Select Drawing File menu window.

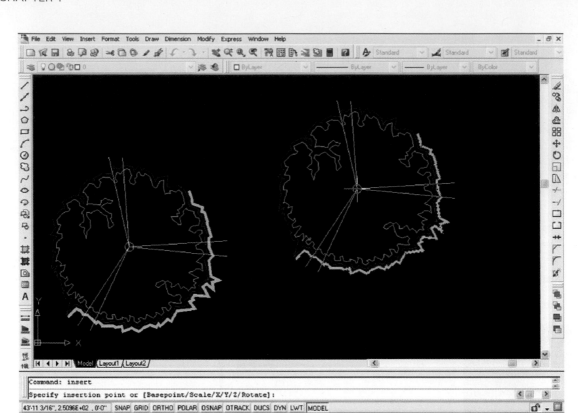

FIGURE 1–70 To set the block in place, just click on the left button of the mouse.

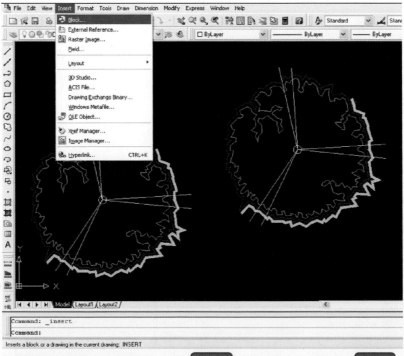

FIGURE 1–71 At the top toolbar under **Insert**, scroll down and select **Block**.

author's note

Block is an object that can be further manipulated by the CAD user. Blocks can be copied, deleted, scaled up/down, or even exploded (turned back to individual objects). Many landscape architects are discovering that they can create an entire library of tree symbols using the block format. Some blocks can be created to serve double duty as "fancy graphics or construction planting symbols" by creating layers for specific function and later turning a specific layer On or Off or choosing Plot/No Plot. In figure 1-72, the QV-Shadow layer is turned off. The key to creating a successful library is to arrange all the block symbols into a matrix table on model space. Then turn this drawing into a standard drawing with the title block frame set up into several page sizes and layout options. This drawing can become the standard drawing on which every single new drawing is based.

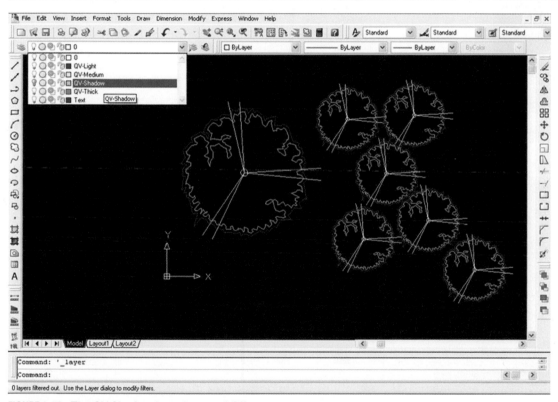

FIGURE 1–72   The QV-Shadow layer is turned Off.

Hatching and Gradient Tool

The Hatching tool will enable the CAD user to insert patterns within the boundaries of selected enclosed polygon or polyline. This tool is mostly used to create hatches to illustrate pavement materials such as brick, tile, or stone, and types of planting such as groundcovers and/or flower beds. This tool requires an enclosed polyline or polygon to work. For the purpose of illustrating this tool, the authors created a simple rectangle using the rectangle tool.

Enable the Hatch tool by left-clicking on the tool icon. This action will launch the **Hatch and Gradient** pop-up menu window (figure 1-73). Under Type and Pattern, the CAD user has the option to select a Predefined (Default CAD), User Defined, or Custom pattern type. Select Predefined to allow selecting pattern types from the default CAD pattern selection. Under this, the scroll Pattern window shows a list of pattern names to select from, or clicking on the button next to the scroll window will launch the **Hatch Pattern Palette** pop-up menu window (figure 1-74). This pop-up menu window shows the images of the pattern options available from the Predefined CAD default selection.

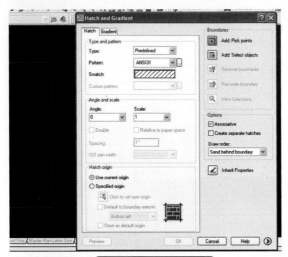

FIGURE 1–73 The **Hatch and Gradient** pop-up menu window.

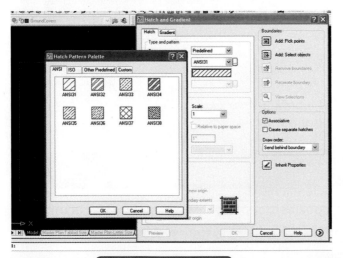

FIGURE 1–74 The **Hatch Pattern Palette** pop-up menu window.

For the purpose of illustrating this tool, the authors selected from the Other Predefined tab the Brick option. This action will close the **Hatch Pattern Palette** pop-up menu window, thus returning to the previous window. Please note that the Swath window now shows the image of the selected hatch pattern. The next action is to define the angle and scale of the pattern. The degree of the angle will establish the orientation of the main axis of the pattern. The authors specified 45° for this demonstration. Next is the scale of the hatch pattern. The scale will change the size of the repetitive modules (bricks) that made the pattern. Because the authors selected brick as the pattern, by logical deduction, the scale of the pattern must be equal to the size of a brick. The standard brick size is $\pm 3.5' \times 8''$ so the scale of the pattern matches these dimensions. Since there is no formula to match the scale of the pattern with the dimensions of the brick, an academic guess must suffice for the moment. The authors typed 10 on the scale as an experimental guess. The next step is to select the Boundaries of the hatching area. There are two methods: Add—Pick Point and Add—Selected Object. The Add—Pick Point will allow the CAD user to click (left button) in the center of the boundaries of the object only if the object is an enclosed-joint polyline/polygon. The Add—Selected Object will let the CAD user select the boundaries by clicking on one or several sides/lines of the enclosed objects. Left-click on the Add—Select Object. This action will hide the **Hatch and Gradient** pop-up menu window; then left-click on one of the lines of the object, and right-click on the mouse to select enter (or hit Enter). This action will return you to the **Hatch and Gradient** pop-up menu; now hit the OK button. This will turn off the pop-up menu window and a hatch pattern will appear within the boundaries of the object selected (figure 1-75).

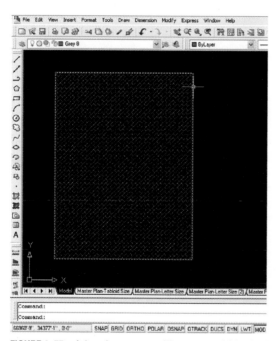

FIGURE 1–75 A hatch pattern will appear within the boundaries of the object selected.

To double check the scale of the hatch pattern, the authors suggest drawing a rectangle with the dimensions of the standard brick pattern and then rotating the rectangle to match the angle of the hatch pattern to compare (figure 1-76).

Figure 1-76 shows that the scale of the pattern is too small to match the actual size of the brick. To change the scale of the pattern, without deleting the hatch pattern and to repeat the process with a different scale value, select-enable the **Properties Palette** (figures 1-48 and 1-49). This will launch the **Properties Palette** pop-up menu. Scroll down until Pattern is listed (figure 1-77). Note that Type, Pattern Name, Angle, and Scale are listed under Pattern. Click inside the Scale scroll window and retype a different number/value and hit Enter. The scale of the hatch will change according to the new value. Continue changing the scale until the pattern matches the size of the brick drawn at scale for comparison. Once the approximate scale that matches the actual size of the brick is reached, hit Enter, close the **Properties Palette** by left-clicking on the X mark, and delete the reference brick-rectangle.

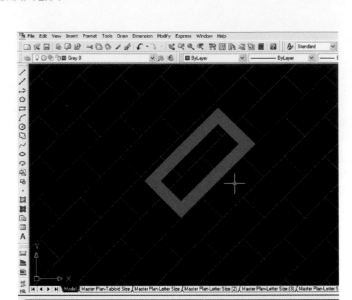

FIGURE 1–76 The dimensions of the standard brick pattern against the hatch pattern.

FIGURE 1–77 The **Properties Palette** pop-up menu.

The **Properties Palette** will also enable the CAD user to change the existing hatches type by changing the appropriate values on the Properties window, without the need of rehatching the areas.

The Gradient tool will enable the CAD user to render an area using a shade of one color or the transition between two colors. This tool will grade the color from light to dark values, giving a reflective quality to the selected object.

This tool requires an enclosed polyline or polygon to work. For the purpose of illustrating this tool, the authors created a simple rectangle using the rectangle tool. Enable the Gradient tool by left-clicking on the tool icon. This action will launch the **Hatch and Gradient** pop-up menu window (figure 1-78). Under Color, the CAD user has the option to select One Color or Two Color combinations. Next to the color preview window (Blue is the default color) there is a button that will launch the **Select Color** pop-up menu window (figure 1-79). This window will enable the CAD user to change the color per **Index Color**, True Color (Default option), and Color Books. Select your color preference. The authors have used the default blue color for this demonstration. Hit the OK button to confirm/set your color selection. This action will return you to the previous pop-up menu window. Under your color preview window, there are six different gradient selections. Choose any for this demonstration.

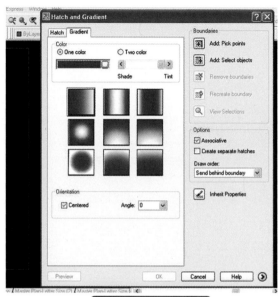

FIGURE 1–78 The **Hatch and Gradient** pop-up menu window.

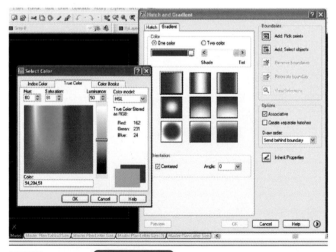

FIGURE 1–79 The **Select Color** pop-up menu window.

Under Orientation, select Centered or Angle per user needs. The next step is to select the Boundaries of the gradient area. There are two methods: Add-Pick Point and Add-Selected Object. The Add-Pick Point will allow the CAD user to click (left button) in the center of the boundaries of the object only if the object is an enclosed-joint polyline/polygon. The Add-Selected Object will let the CAD user select the boundaries by clicking on one or several sides/lines of the enclosed objects. Left-click on the Add-Select Object. This action will hide the [Hatch and Gradient] pop-up menu window; now left-click on one of the lines of the object and right-click to select enter (or hit Enter). This action will return you to the [Hatch and Gradient] pop-up menu, and then hit the OK button. This will turn off the pop-up menu window and a gradient pattern will appear within the boundaries of the object selected (figure 1-80).

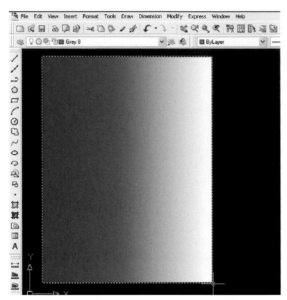

FIGURE 1–80 A gradient pattern will appear within the boundaries of the object selected.

author's Note

Per the author's experience, the gradient tool will consume a lot of memory, making the CAD drawing extremely large. In landscape architecture practice, the gradient tool is used to create background and texture to enhance the graphics. Use with extreme caution.

Leader-Label Lines Text and Dimension Lines

Dimension Lines are commonly used to illustrate an object's reference location, measurements, size, and proportions on a stake-out plan or construction detail drawings. The Dimension Linetypes are located on the top of the toolbar, under [Dimension], and they are Quick Dimension, [Linear], Aligned, Arc Length, Ordinate, Radius, Jogged, Diameter, Angular, Baseline, Continue, Leader, [Tolerance], Center Mark, Oblique, and Align Text (figure 1-81). Also in the same pop-up menu window are listed [Dimension Style], [Override], Update, and Re-associate Dimensions.

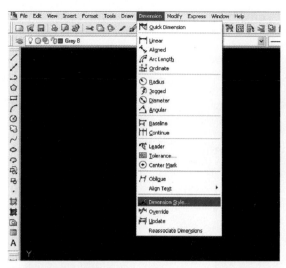

FIGURE 1–81 The Dimension Linetypes.

These dimension type options are the most common dimension lines used for documenting proportions, size, measurements, and reference locations of the objects drafted on the screen. To enable any one of these tools, left-click on the tool icon. The authors suggest creating a simple rectangle and a circle of any size in model space for illustrating this tool. Next, turn the Osnap option on (figure 1-46), and make sure the Endpoints, Intersection, and Center are checked. Then select **Linear** dimension type (figure 1-82). This action will return the screen to model space; now left-click to set the first point of the linear dimension, followed by a left click on the mouse to set the second dimension point (figure 1-83). A third left click of the mouse will set the dimension line in place (figure 1-84).

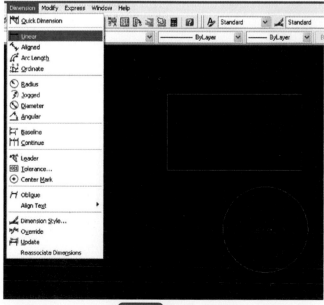

FIGURE 1–82 Selecting **Linear** dimension type.

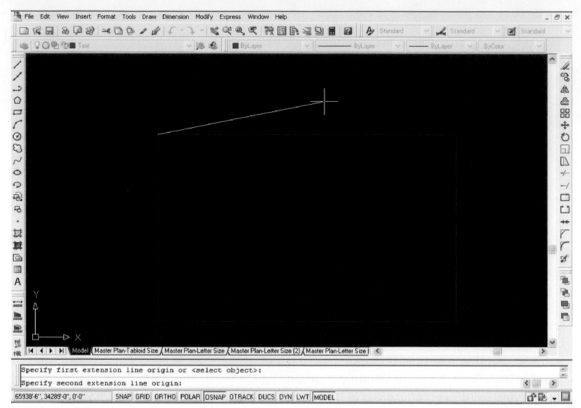

FIGURE 1–83 Setting the first point of the linear dimension.

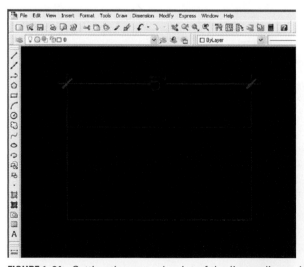

FIGURE 1–84 Setting the second point of the linear dimension.

To create a dimension line (radius) for the circle, select Radius dimension type (figure 1-85). This action will return the screen to model space; left-click on the circle, followed by a left click on the mouse to set the radius dimension in place (figure 1-86).

FIGURE 1–85   Select Radius dimension type.

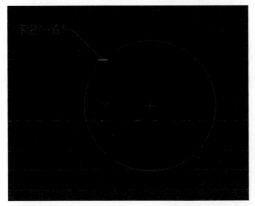

FIGURE 1–86   A left click on the mouse to set the radius dimension in place.

The Leader tool will allow the CAD user to create a label line with text. To create a label line with text, select **Multileader** dimension type (figure 1-87). On previous CAD editions, **Multileader** is the same as leader (figure 1-85). This action will return the screen to model space; left-click on the object to be labeled, followed by a left click on the mouse to set the second point of the arrow line (figure 1-88). A third left click will set the third point or second segment of the label line, but make sure the Ortho mode is set On (briefly) so that the second line segment will be horizontal. Notice that the Command Prompt window now reads "Specify text Width." Ignore for the moment and hit Enter; this is followed by a new Command Prompt window message that reads "Enter first line of annotation text." Type the content of the label. In the authors' screen, type the word "Rectangle." A second Command Prompt window message will ask the CAD user for a second line. If no second line is desired, just hit Enter to set the text and label in place (figure 1-89).

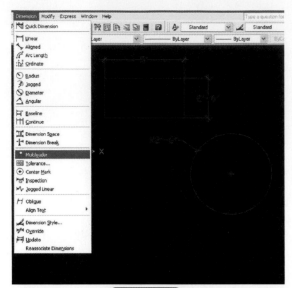

FIGURE 1–87 Under **Dimension**, scroll down and select **Multileader** dimension type.

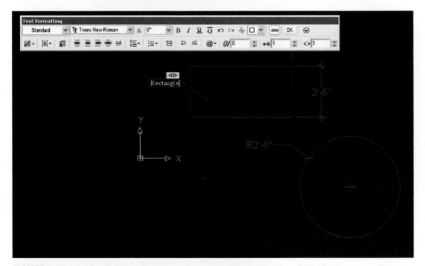

FIGURE 1–88 Left-click on the mouse to set the second point of the arrow line.

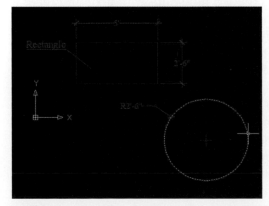

FIGURE 1–89 Hit Enter to set the text and label in place.

author's note

Owing to the introductory nature of this book, the authors cannot include in this edition all the different types of dimension types. The authors recommend that CAD users experiment with all the dimension types listed under **Dimension** in the top toolbar. This will help the CAD user to become familiar with the specific use of each one of the dimension types listed.

The **Dimension** types illustrated above are the default style option set by CAD. To change the style of the dimension lines, under **Dimension**, scroll down and select **Dimension Style** (figure 1-90). This will launch the **Dimension Style Manager** pop-up menu window (figure 1-91).

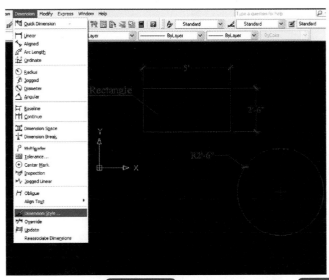

FIGURE 1–90 Under **Dimension**, scroll down and select **Dimension Style**.

FIGURE 1–91 The **Dimension Style Manager** pop-up menu window.

The **Dimension Style Manager** will enable the CAD user to modify the dimension styles to suit the user's preferences. The default and current CAD dimension style is listed in this window as "Standard." The CAD user can modify the "Standard" style by hitting the **Modify** button or clicking on **New**. The authors recommend creating a "New" dimension style and leaving the Standard format intact for later comparison. Left-clicking on the **New** button will launch the **Create New Dimension Style** pop-up menu window. The default option of CAD is listing the new style as "Copy of Standard." Relabel the name by clicking inside the window and typing your new dimension style name (figure 1-92). This will launch the **New Dimension Style** : **My Dimension Style** (name given by authors) pop-up menu window (figure 1-93).

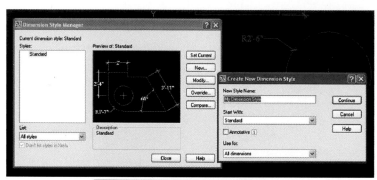

FIGURE 1–92 The **Create New Dimension Style** pop-up menu window.

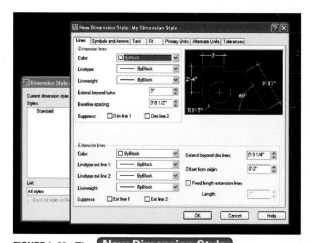

FIGURE 1–93 The **New Dimension Style** pop-up menu window.

Inside the **New Dimension Style** pop-up menu window, the CAD user will find the following Tab options: **Lines**, **Symbols and Arrows**, **Text**, **Fit**, **Primary Units**, **Alternate Units**, and **Tolerance**.

By selecting the **Lines** tab of the **New Dimension Style**, the CAD user can modify the Dimension Lines and Extension Lines, Color, **Linetype**, **Lineweight**, Extend Beyond Ticks, and Baseline spacing by clicking on the scroll arrow options or typing the size preference (figure 1-93).

Under the Symbols and Arrows tab (figure 1-94), the CAD user can change the Arrowheads, First-second type, Leader type, Arrow size, Center mark options, Arc length symbol options, and the Radius dimension jog angle.

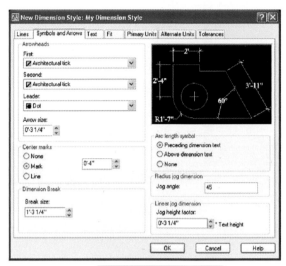

FIGURE 1–94 The Symbols and Arrows tab.

The Text tab will let the CAD user change the text style, text color, fill color (the authors recommend that the None option on the Fill Color be selected to prevent the fill color screening/blocking the text on the finish plotted drawing), text height, fraction height scale, text placement, and text alignment of the Text Appearance options. To change the text font, click on the button next to Text Style . This will launch the Text Style pop-up menu window (figures 1-95 and 1-96).

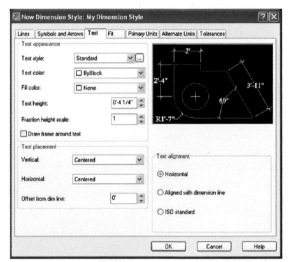

FIGURE 1–95 The Text tab under the New Dimension Style pop-up menu window.

FIGURE 1–96 The **Text Style** pop-up menu window.

Modify—Select the font style, size, and format per user needs, then hit the Apply button to set the changes in place and returning to the previous pop-up menu window.

The **Fit** tab will provide further options to customize the placement of the text and arrows inside the dimension lines (figure 1-97).

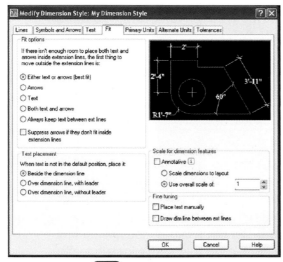

FIGURE 1–97 The **Fit** tab.

The **Primary Units** tab will let the CAD user set the Units—Measurement system of the dimension lines (figure 1-98). **Alternate Units** will also let the CAD user specify a second Units system for the dimension lines (figure 1-99). Setting the **Tolerance** tab to a specific method option will set the automatic format of placement of the text in a case were there is no space to fit the text in place (figure 1-100). The authors recommend leaving this option as None since, once the dimension lines are set in place, the CAD user can click on them and move them around to fit.

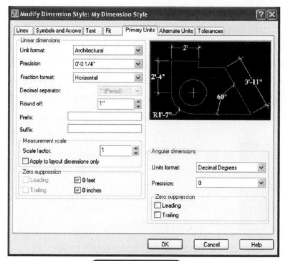

FIGURE 1–98   The **Primary Units** tab.

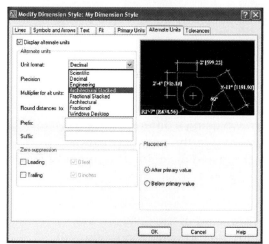

FIGURE 1–99   The **Alternate Units** tab.

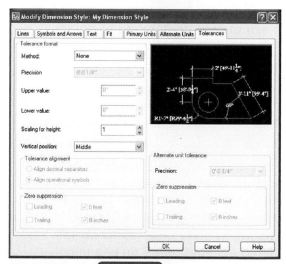

FIGURE 1–100   The **Tolerance** tab.

Once the CAD user has set all tabs to their preference options, the OK button will have to be hit to set these options. This action will return you to the previous pop-up menu window where now My Dimension Style is listed as an option. Click on the My Dimension Style (highlight in blue color), and then hit the Set Current button to make this dimension style the current system (figure 1-101). Please note that Current Dimstyle on this window now states My Dimension Style. After this is confirmed, hit the close button to close all pop-up menu windows and return to model space.

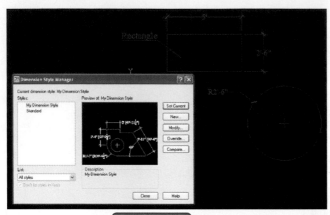

FIGURE 1–101 Hit the Set Current button to make this dimension style the current system.

Please note that the changes to the Dimension Style do not change the pre-existing dimension lines on the drawing since they were drawn using a different Dimension Style. To change the dimension lines on the existing drawing to the new My Dimension Style, select all dimension lines by left-clicking on the screen and dragging the cursor to create the selection green field over all dimension lines (figure 1-102).

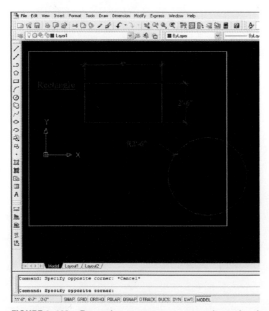

FIGURE 1–102 Drag the cursor to create the selection green field over all dimension lines.

Next, launch the **Styles** pop-up menu window by placing the cursor arrow on the top tools and right-click (figure 1-103). There, scroll down and select **Styles**. This will open the **Styles** pop-up tool window (figure 1-104). In the center window, click on the scroll arrow and select from the list **My Dimension Style**. This will change all the selected dimension lines from the Standard to the new **My Dimension Style** format (figure 1-105).

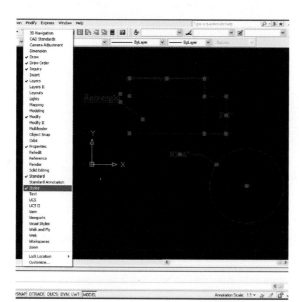

FIGURE 1–103 Scroll down to select **Styles** option.

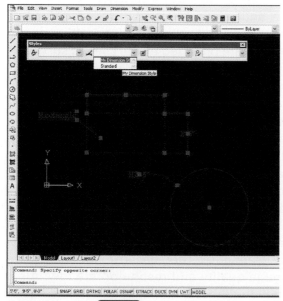

FIGURE 1–104 The **Styles** pop-up menu window.

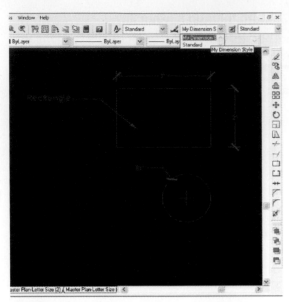

FIGURE 1–105 Change all the selected dimension lines from the Standard to the new My Dimension Style format.

The Styles window can be closed by clicking on the X mark or place-docked in the top toolbar for future reference (figure 1-105).

Inserting a Scanned Image

Although CAD is not the best desktop publishing image manipulation software, it is designed to be compatible to accept and manipulate images to a certain extent. Images can be inserted in CAD model or paper space, and can be enlarged or reduced, moved, rotated, stretched, zoomed in and zoomed out, used as a background for drafting on top of it, and also as a base plan. In theory, base plans are provided by the client's surveyor-engineer-architect in a digital form (CAD plan or image). Per authors' professional practice experience, the base plan is usually handed in by the client as a simple photocopy of a plat drawing, done by a surveyor or civil engineer, and containing the properties' boundaries, topographic information, structures, and any other information. Depending on the size and complexity of the site, the client may provide a CAD survey drawing, but in the event that this format is not available, a photocopy of a plat will suffix. This plat photocopy, as long as it drawn at scale and contains a north arrow, can be scanned (digitized) as a TIFF (Tagged Image File Format) or JPEG (Joint Photographic Experts Group) format, which in turn can be inserted on CAD, so it can be used as a base image/plan to trace over with CAD. The TIFF and JPEG formats are the recommended image format to be inserted in a CAD drawing because the images are compressed into a more manageable memory (bytes) size. CAD users noted that, because the images are compressed into a small size format, the image quality and resolution are low. Also, there is a low percentage of image distortion in the process of transferring information from print to digital media. A ready CAD survey format is preferred, but at the early stages of the design process (not technical, construction drafting) is acceptable up to certain degree.

Inserting a scanned image into a CAD drawing has become a common practice in the landscape architecture since it provides a drawing use as an image/base to trace over using CAD lines. To insert an image into CAD, the authors recommend that the scanned image format be TIFF or JPEG, a layer named Image be created and set as the active layer, and the image be inserted in model space. The next step is to select **Insert** on the top toolbar, scroll down, and select **Raster Image** (figure 1-106). This action will launch the **Select Image File** pop-up menu window. Browse in this window to locate the image, select the image so it will show in the File Name window, and then hit the Open button (figure 1-107).

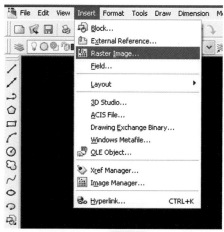

FIGURE 1–106 Under **Insert**, scroll down to select **Raster Image**.

FIGURE 1–107 The **Select Image File** pop-up menu window.

Hitting the Open button will launch the **Image** pop-up menu window. Leave the Path type as Full path. Under this scroll window you will find Insertion Point, Scale, and Rotation. Leaving the green checkmark "On" in the "Specify on-screen" options will translate as being defined by the CAD user directly on the screen. Deselecting the green checkmark in the "Specify on-screen" option will let the CAD user define these items exactly before the image is inserted on the drawing. The authors recommend leaving these items with the green check "On" so the image can be inserted at any point on the screen and later stretched to fit to size/scale (figure 1-108). After completing your selection, hit the OK button.

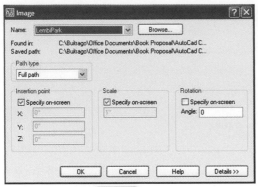

FIGURE 1–108 The **Image** pop-up menu window.

Hitting the OK button on the previous **Image** pop-up menu window will return the drawing to model space. Please note that the cursor arrow will become the insertion point or first corner of the image. One click with the left button of the mouse will set the first corner of the image. Dragging and stretching with the cursor will make a rectangle appear on the screen. This rectangle is the preview of the edges and limits (frame) of the image. Stretch the rectangle so it fits the screen and then left-click to set in place (figure 1-109). The image will now appear within the boundaries of the rectangle (figure 1-110).

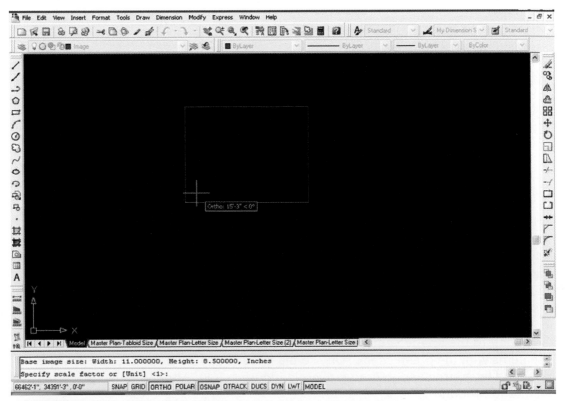

FIGURE 1–109 The image rectangle/frame.

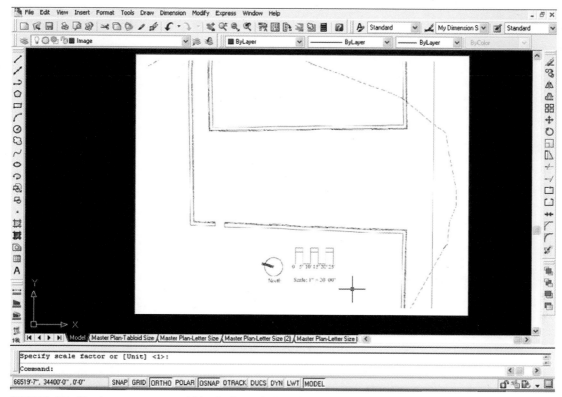

FIGURE 1–110 The image appears within the boundaries of the rectangular frame.

Because the inserted image contains a graphic scale and a north arrow, the CAD user can rotate the entire image to match the true north orientation of the CAD drawing (90° vertical), and scale the drawing up or down to match the image to a reference distance.

Rotating the drawing is done simply by using the Rotate tool (figure 1-40). To stretch/scale the drawing, you must first draw a reference line or object at a specific distance (measurements) that matches the measurements of a specific line or object of the original image. This reference point will become the point to which the image will be stretched to match. To stretch the image, simply left-click on the image frame (rectangle) and on any of the endpoints of the frame. Drag and stretch or reduce the image frame until the reference line or object of the image appears to match the drawn reference line/object. Because this is a visual approach, the scale of the image will not be 100% accurate, but close enough to commence drafting a base plan.

CAD users are aware that on certain occasions the image may go missing from the screen and show up as an empty rectangle with error message text in its place. This is because the image is not permanently attached to the drawing. What is attached to the drawing is the address of where the images are originally located. The address location may have been corrupted if the original image was deleted from the file, the file location was changed, or even the file name was changed. To restore the address (path) of the image, look under **Insert** on the top of the toolbar, scroll down, and select **Image Manager** (figure 1-111). This will launch the **External Reference** pop-up menu window (figure 1-112).

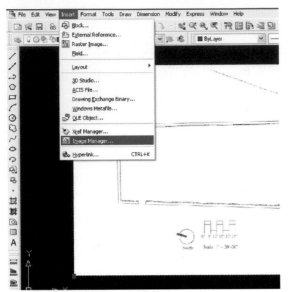

FIGURE 1–111 Under the **Insert** top tab, scroll down and select **Image Manager**.

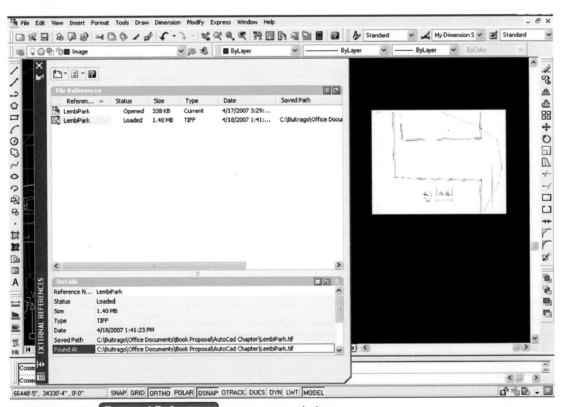

FIGURE 1–112 The **External Reference** pop-up menu window.

In this window, under File Reference, select the image file, name of the missing image. Next look under Details, and left-click inside the Found At window. This will show the current address (path) where the image was originally located. To change the path, click on the button located at the end of the Found At window. This will launch the

Select Image File pop-up menu window (figure 1-113). Using the browsing buttons, locate the missing image file, then make sure it is listed in the File Name window and hit the OK button.

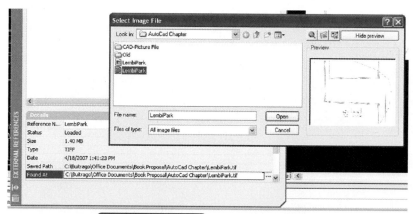

FIGURE 1–113 The **Select Image File** pop-up menu window.

After hitting the OK button, the **Select Image File** pop-up menu window will close. This will restore the image location address (path), thus showing the image on the screen. The **External Reference** pop-up menu window can be closed by left-clicking on the X mark.

Setting the Title Block Page

At this point, the readers should be capable of drafting a simple drawing. Simple shapes or objects will do at this stage. As the user progress on their CAD exploration and practice, more complex form will evolve naturally. Start with familiar objects, perhaps a piece of furniture such as a chair. Draw this chair in model space and create different drawings or views of this object as if you were looking at it from the top (plan) or the side (elevation). Also, create different layers to isolate different elements of the chair such as metal or leather. Use the layers to create different linetypes, lineweights, and arrange in a color-by-layer code system to visually differentiate materials. Assume this chair is placed on an outdoor terrace. Create different layers to add/draw other elements such as planting and hardscape on plan view. Follow the previous instruction on how to create planting symbols as Blocks. Use the Hatch tool to add a brick surface on which the chair is sitting. Practice makes perfect.

The authors created a simple, hypothetical pocket garden plan called **Lembi Park** (figure 1-114). This pocket garden will be used through this book to illustrate several different rendering techniques and software applications. The authors created Lembi Park on model space and used architectural units. This plan contains several planting symbols (on a Write Block format), hatch area representing different hardscape elements such as brick, pavers, or ground covers, and different linetypes to represent building or hardscape elements; all drafted elements were arranged by layer properties. Your drawing does not need to be as complex as Lembi Park, but your plan must be drawn on model scale and using architectural units. It must also contain several objects drawn on different colors, linetype, and lineweights. Include several areas containing a hatch pattern. Please refer to the "Creating your first drawing" section of this chapter for further information. Once your CAD rendering on model space is complete, the next step is to create a page layout with the standard title block information.

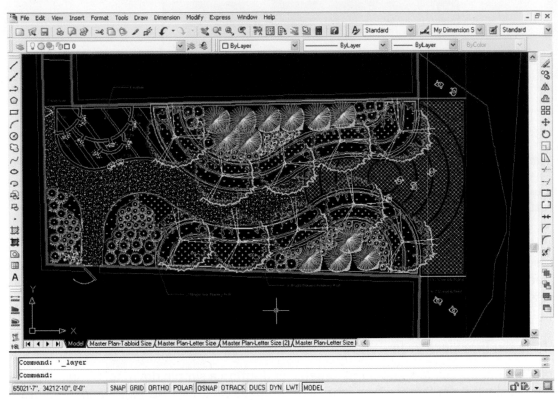

FIGURE 1–114 The authors' CAD plan for Lembi Park.

In order to create a Page Layout, at the bottom toolbar, hit the tab that reads Layout 1. This will close the model space view and turn the drawing into paper space (Layout 1) view. Earlier in this chapter, the authors explained how to set up the limits of the Drawing (Paper Size) by selecting the **Page Setup Manager** under **File** — **Page Setup Manager** on the top toolbar of the screen (figure 1-115). Select the paper size and orientation per your plotter/printer options. The authors have used a standard tabloid size format (11″ × 17″) on landscape view for illustrating this section. Once your paper size and orientation are selected, the drawing drafted in model space should be displayed inside a rectangle at the center of a white page background (figure 1-116). The frame where the drawing is located is called viewport.

On paper space, the authors recommend that you type your basic Title block information such as Title of Project, Address, Client, Your Company Logo, License Number and Registered Landscape Architect Stamp, North Arrow, Scale, Frames, and any other information as needed. The authors also recommend that this information be drawn on a specific layer. The rectangle (**Viewports**) will be printed if the layer where the **Viewports** is located is set to plot. Depending on the CAD user preference, such as the author's choice, the **Viewports** (rectangle) has been set up to a nonplot layer. Setting the **Viewports** (rectangle) to a nonplot layer would not affect the content of the **Viewports** ,

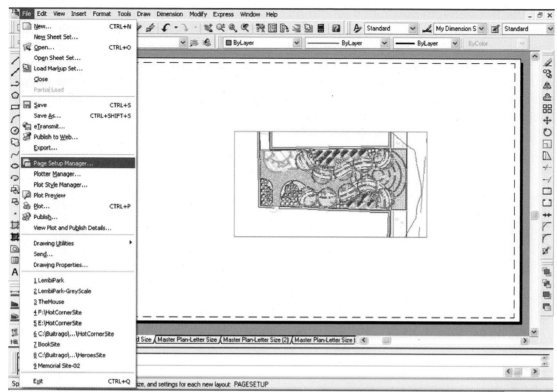

FIGURE 1–115 Under the **File** top tab, scroll down and select **Page Setup Manager**.

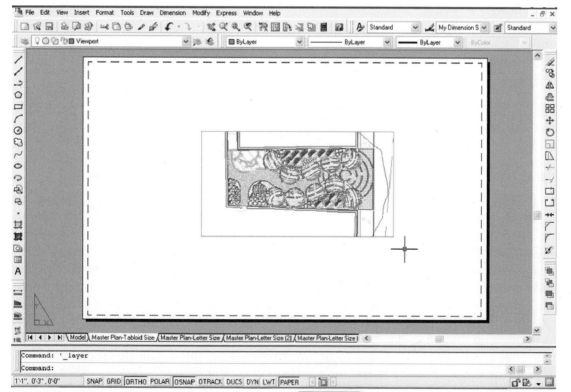

FIGURE 1–116 Paper space view of Lembi Park inside the **Viewports** window.

and thus the drawing contained within its boundaries will be printed. CAD users are aware that on paper space, the limits of your drawing are now set by the limits of the size of the paper. As previously explained in this chapter, there are no limits on model space. In paper space your drawing is confined to the size of the paper, so the dimensions of any object or line drawn in paper space mode are constrained by the edges/ boundaries of the paper size. It is wise to treat paper space as a real piece of paper on which we can draw things as they appear on paper (figure 1-117).

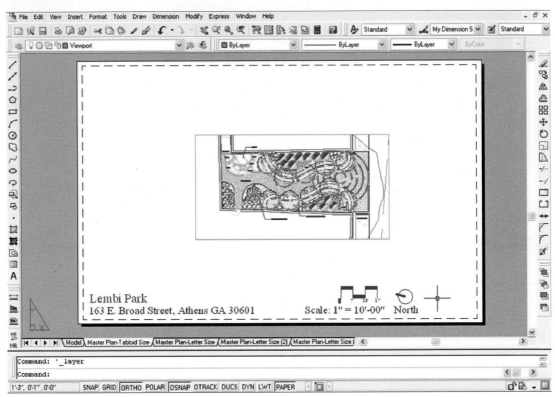

FIGURE 1–117 The author's Lembi Park title block page on paper space.

Setting the Scale of the Drawing with the Viewport Tool

The **Viewports** tool will let the CAD user change the scale and also reveal portions of the drawing on paper space mode. The concept of the **Viewports** can be easily understood with the following analogy. Model space is one sheet of paper without any limits. Paper space is a second sheet of paper that has limits and lies on top of the model sheet of paper. In order to see what lies under the paper space sheet, we must cut through the paper and create a window (**Viewports**). This window reveals what is under the paper space sheet, revealing the objects and lines that are on model space sheet. If we stretch or reduce the **Viewports** window horizontally in any direction, we can frame portions of the drawing that lie under it. Also, if we raise the paper space sheet above (zoom out) the model space sheet, a larger area of the drawing is revealed. The closer (zoom in) the paper space window is, the larger the objects will appear.

To change the scale (zoom factor) of the **Viewports**, open the **Viewports** tool. The **Viewports** tool can be opened by placing the cursor arrow on the top of the toolbar and right-clicking the mouse. This will launch the **Custom** tool pop-up menu scroll window. Scroll down and select **Viewports** by left-clicking on the mouse (figure 1-118). This will launch the **Viewports** pop-up menu box (figure 1-119).

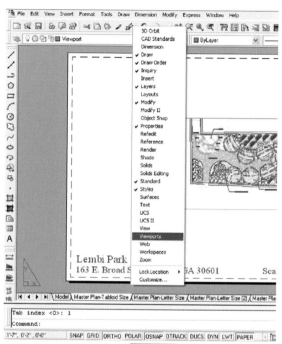

FIGURE 1–118 Under the **Custom** tool pop-up menu, scroll down and select **Viewports**.

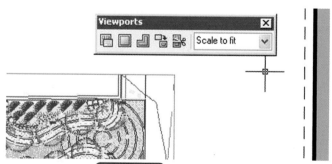

FIGURE 1–119 The **Viewports** pop-up menu window.

To change the scale, left-click on the **Viewports** frame—rectangle, which now will turn into a dashed line, and the endpoints (blue squares) of the frame will appear. Next, left-click on the scroll arrow of the **Viewports** menu box, and select (left-click) from the scroll-down list any of the architects' (fractions) scales (figure 1-120). The author's drawing scale choice is 3/32″ = 1′-00″. This will change the scale of the **Viewports** window (figure 1-121).

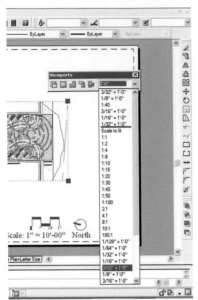

FIGURE 1–120 Drawing scale scroll-down menu options under the [Viewports] menu window.

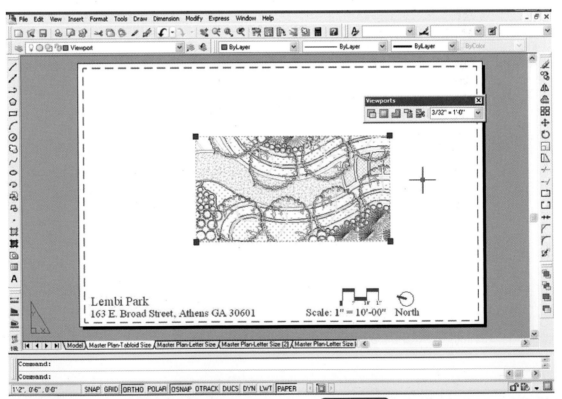

FIGURE 1–121 Click, select, and drag the square corners of the [Viewports] window to enlarge or reduce the [Viewports] to fit the drawing.

To fit the drawing to the [Viewports] window, left-click on the endpoints of the frame, drag the cursor arrow to enlarge or reduce the size, and left-click to set in place. Then use the move tool to move the [Viewports] to fit to paper (figure 1-122).

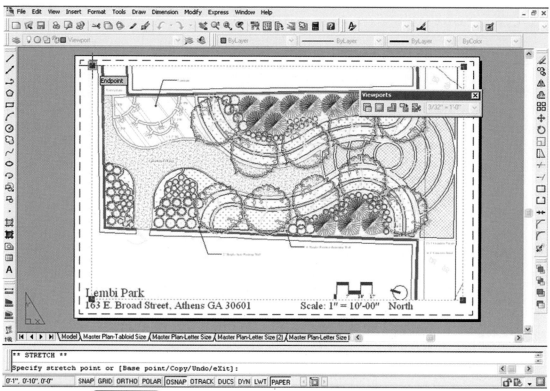

FIGURE 1–122 The **Viewports** set to fit the paper space dimensions.

---

author's note

CAD users are aware that because this drawing was created in the Architectural Length Type, the fraction scales of the **Viewports** tool box scroll menu will work correctly. The scale ratios (1:10, 1:20, 1:30, 1:100, etc.) of the same scroll menu will not work. The scale ratios only work for drawings where the units system was set up for Engineering Length Type. Please refer to the Units section of this chapter. To change the scale of the **Viewports** to traditional Engineer ruler scale (1″ = 10′-00″, 1″ = 20′-00″, 1″ = 100′-00″, etc.), the scale zoom factor tool must be used.

The scale zoom factor works in multiples of 12 units. So if the desired scale is 1″ = 10′-00″, the mathematical formula is 10 × 12 = 120, and thus the scale zoom factor will be 1/120 XP (exponential zooming factor). If the preferred scale is 1″ = 20′-00″, the mathematical formula is 20 × 12 = 240, and thus the scale zoom factor will be 1/240 XP. If the preferred scale is 1″ = 30′-00″, the mathematical formula is 30 × 12 = 360, and thus the scale zoom factor will be 1/360 XP. Last, if the preferred scale now is 1″ = 100′-00″, the mathematical formula is 100 × 12 = 1200, and thus the scale zoom factor will be 1/1200 XP, and so on.

Linetype Scale Command—ltscale

The Linetype Scale command will let the CAD user set the global linetype scale factor in the ⬤ Viewports ⬤ window of the drawing. In essence, this command is used to set the size and spacing per linetype. For example, the spacing of any type of dash line can be set as a large or smaller space between dashed lines by using this command. To engage this tool, first make sure the ⬤ Viewports ⬤ is active (bold lines), then type "ltscale" in the Command Prompt window, and hit Enter (figure 1-123). The Command Prompt window now reads "Enter new linetype scale factor." Enter the number value per user preference and hit Enter. Changing the linetype global scale factor will cause the drawing to regenerate.

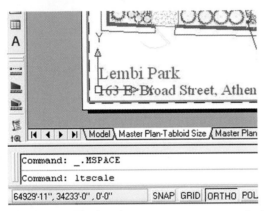

FIGURE 1–123   The ltscale type prompt command window.

CAD users are aware that there is no specific linetype scale factor or ratio that works best for each specific ⬤ Viewports ⬤ scale. The linetype scale number varies per CAD user preference or graphic style. Try different scale values such as 0.0001 through 2.0, and so on, until the desired linetype "look" is achieve.

Locking the Viewport

Per author's teaching experience, one of the most common errors performed by first-time CAD users is omitting to lock the scale of the ⬤ Viewports ⬤ window. This simple task will prevent changing the scale of the drawing by accidental zooming in/out or panning (Hand tool) while the ⬤ Viewports ⬤ window is active (bold line frame). To lock the ⬤ Viewports ⬤ window, left-click once on the ⬤ Viewports ⬤ frame. This action will turn the ⬤ Viewports ⬤ frame into a dashed line. Then select the Properties tool by left-clicking on the tool icon (figure 1-48). This will launch the ⬤ Properties Palette ⬤ pop-up menu window (figure 1-124).

Scroll down until the Miscellaneous Option appears. Under this menu option, left-click on the Display Locked option and select Yes. This will lock the ⬤ Viewports ⬤ window, thus preventing changing the scale by accident. To unlock the ⬤ Viewports ⬤ window, just open the ⬤ Properties Palette ⬤ window, select the ⬤ Viewports ⬤ once, and under the Display Locked option, select No.

Under the Display Locked option is the Standard Scale options. Like the ⬤ Viewports ⬤ window tool options (figure 1-119), the scale of the drawing can be set here as well. Make sure the Display Locked option is No (unlock) in order to be able to access and change the scale of the ⬤ Viewports ⬤ under the ⬤ Properties Palette ⬤ pop-up menu window options.

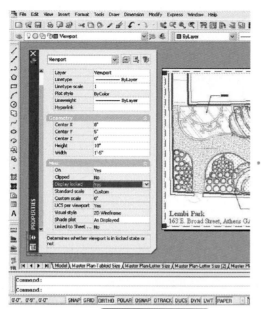

FIGURE 1–124   The Properties Palette pop-up menu window.

Saving and Printing

Saving your document at different stages of the drawing is highly recommended. Depending on the CAD user hardware reliability, saving your work in progress will prevent loosing any data by a system crash or failure due to hardware malfunction, computer viruses, severe weather, Internet traffic connection, memory capacity, and others. To save a document, look under File at the top of the toolbar, scroll down and select save. The Save command works on the same standardized computer software format. The pop-up menu window will require the CAD user to give a file name and location to save the document. Just follow the pop-up menu windows accordingly.

To print the drawing, look under File at the top of the toolbar and select Plot. This will launch the Plot pop-up menu window (figure 1-125). Make sure that the Printer/Plotter, Paper Size, Plot Style Table (Pen Assignments), Plot Area (Layout), Plot Scale (1:1), and Drawing Orientation are set as per user preference. Hit the Preview button to get an image of your Plot-printed drawing (figure 1-126).

If the preview window shows the document set as per user preference, just hit the Print button icon located on the top of the toolbar of the preview window. Also hitting ESC will return the user to the previous window; here, hitting the OK button will print the document.

author's note

CAD users are aware that the printer settings of the local desktop will always override the printer settings of the AutoCAD software or any active software printing settings. Make sure that, before launching any application, the local printer setting of your desktop is set to user preference.

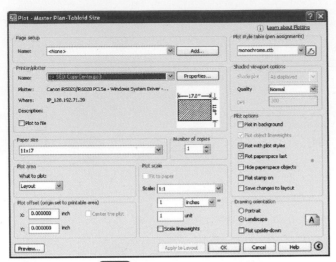

FIGURE 1–125   The **Plot** pop-up menu window.

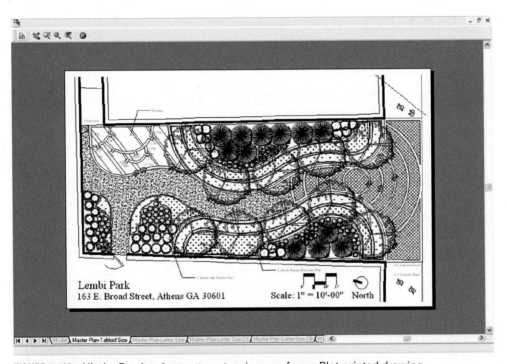

FIGURE 1–126   Hit the Preview button to get an image of your Plot-printed drawing.

Print PDF and E-mail to a Service Provider

Saving or plotting your drawing as a Portable Document Format (**PDF**) is becoming the easy way to e-mail your CAD drawing document. The PDF format was first introduced in 1993 by Adobe Systems to provide an easy format that is readable by the most common desktop software used in an office environment. This format also compressed the large memory size (bytes) of the document, making it easy for electronic transmission (e-mail).

Also, this format is the most compatible with all kinds of printer devices, making it easy to plot regardless of CAD printer system requirement.

Readers, please refer to the third chapter of this book for further instructions on how to import a CAD drawing into Adobe Photoshop CS2 as a PDF format.

TERMS

AutoDesk—A San Rafael, California, based company that in 1982 created a computer-generated drafting software that is known as AutoCAD.

AutoCAD, CADD, or **CAD**—Acronyms for computer-aided design and drafting.

DWG—Abbreviation of drawing.

Lembi Park—A hypothetical site used by the authors for demonstration purposes only.

PDF—The acronym for Portable Document Format.

SketchUp

By Professor Ashley Calabria

CHAPTER OBJECTIVE

This chapter introduces you to SketchUp's three-dimensional drawing techniques. By the end of this chapter you should be able to model a plan-view drawing in 3D, apply different effects depending on end-product objectives, and prepare the document for a hard copy or a video fly-through.

Introduction to SketchUp

"Founded in 1999 and based in Boulder, Colorado, @Last Software the maker of SketchUp was created by a small group of AEC (Architecture Engineering Construction) software industry veterans. This group envisioned developing 3D design software that would make design exploration accessible to everyone.

In the development of SketchUp, @Last Software kept in mind several design principles:

- To allow designers to *draw the way they want* by emulating the feel and freedom of working with pen and paper in a simple yet elegant interface.

- To enable the user to have *fun*.

- For the program to be *easy* to learn and use.

- And to enable designers to *play* with their designs in a way that is not possible with traditional design software.

In March of 2006 @Last Software was acquired by Google, and shortly thereafter released a free version of SketchUp allowing everyone to create models in 3D." http://www.sketchup.com

The current version of Google SketchUp for purchase is SketchUp Pro 6, which is described in this chapter. Google SketchUp 6, the free version, is also applicable for a majority of this chapter. At the time this book was written, the main SketchUp website had a vast array of examples for different design fields, online tutorials, and a fair amount of component downloads. Visit www.sketchup.com after you go through this chapter for more detailed information on the program.

author's notes

- All geometry consists of points, edges, and surfaces.
- There are three **axes** in the drawing. The **red axis** is similar to width, the **green axis** is similar to depth, and the **blue axis** is similar to height.
- Educators can receive an educator's license and the latest version at http://www.sketchup.com
- The SketchUp website has some great interactive tutorials at http://sketchup.google.com

A word of caution: SketchUp is an incredibly easy and fun way to manipulate and visualize designs, but there is much debate on its use for creating a finished graphic product. Although the image looks great on screen, hard copies have a tendency to be less dynamic in line quality, depth perception, and effective texture expression resulting in images that look a bit heavy and flat. The newer addition of Effects in SketchUp has provided a wider variety of end-product simulations that print out better than those from previous versions, but based on interviews with landscape architecture firms who use the program, we find that SketchUp is currently most often used for creating a quick 3D "wire frame" that can be printed and traced over by hand. This chapter, although introductory, goes a bit further by covering materials, effect styles, and animation in anticipation of SketchUp's increasing popularity, ease of use, and newer versions addressing better print-quality graphics.

The SketchUp Screen

Figure 2-1 illustrates the default SketchUp Pro 6 screen that will automatically open. Depending on the SketchUp version you have, your screen might look a little different. If you allow the cursor to hover over each icon for a few seconds, a yellow tag will pop up with the name of the tool. To add toolbars that might be missing, go to **View**—Toolbars and select the ones you need.

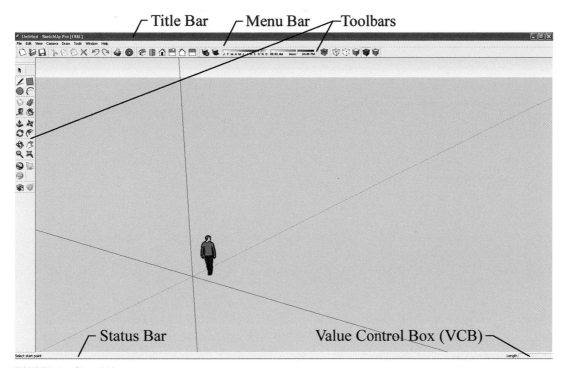

Title Bar   Menu Bar   Toolbars

Status Bar   Value Control Box (VCB)

FIGURE 2–1 SketchUp screen.

On the top of the screen you will see the default Title Bar showing the file path where the drawing is located and the name of the drawing. Below the Title Bar is the Menu Bar and then the Standard toolbar. The most commonly used tools are along the left toolbar. At the bottom of the screen is the Status Bar. On the left side of the Status Bar is an area that will highlight tips and special features of tools. On the right is the **Value Control Box** or **VCB**. The VCB is the area used to enter dimensions of items or to manipulate elements. You do not need to click inside the VCB for SketchUp to read your information—anything you type will go directly into the VCB.

Toolbars

The most commonly used tools in this chapter are labeled in the Toolbar image as seen in figures 2-2 and 2-3 (the figures show a screen shot of the toolbars with labels and leaders pointing to the different tools described and used in this chapter). The tools are described more thoroughly throughout this chapter.

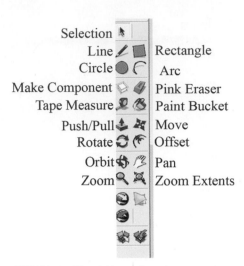

FIGURE 2–2 SketchUp toolbar: Large tool set.

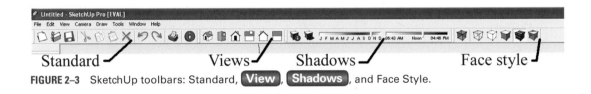

FIGURE 2–3 SketchUp toolbars: Standard, **View**, **Shadows**, and Face Style.

Basic 2D Drawing

We will start by creating a series of vignettes that will allow you to become familiar with the basic drawing principles for producing a project. These skills can then be applied to the existing AutoCAD project.

Units

The default SketchUp units are architectural. To change units for future projects go to **Window** — **Model Info** and select Units from the left hand list. Then select the units and precision that you want SketchUp to read. This tutorial will use architectural units, so you do not need to change the units.

Zoom and Pan

Zoom in or out using the scroll wheel on the mouse or select the Zoom Extents tool, which will allow you to view your full drawing area. Pan around your drawing by holding down the scroll wheel and the shift key or select the Pan tool, which will allow you to move around your drawing.

Views

The **Views** toolbar (figure 2-4) allows you to select specific angles for viewing your project.

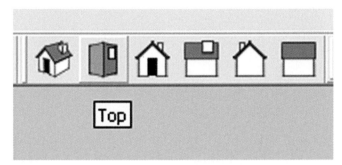

FIGURE 2–4   Isometric, top or plan, front, right, back, and left views.

Line Drawing

Click on Top View from the **Views** toolbar. Activate Line tool, then click in the upper left of the drawing area and release. Move your cursor to the right in the direction of the red axis (figure 2-5). The **Inference Tag** "On Red Axis" will appear, which will assist you in aligning points, lines, surfaces, or axes. Type in 20′. Remember, you do not need to click in the VCB to type—SketchUp will automatically read your information. This will make a line segment 20′ long in the direction of the red axis. Next, move down along the green axis and type in 15′ (figure 2-6).

FIGURE 2–5   Drawing along the red axis.

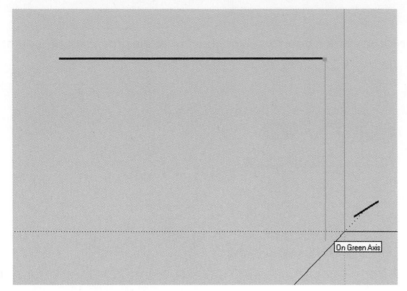

FIGURE 2–6 Drawing along the green axis.

Move back along the red axis, but this time do not type in 20′. When you move back along the red axis toward the beginning point, you will see a dashed green line run from the beginning point to the end of your cursor. This inference line refers to the start point and ensures alignment. Pick that point and then finish by moving the cursor up along the green axis. This closes the box and creates a **Skin** or **Surface** (figure 2-7).

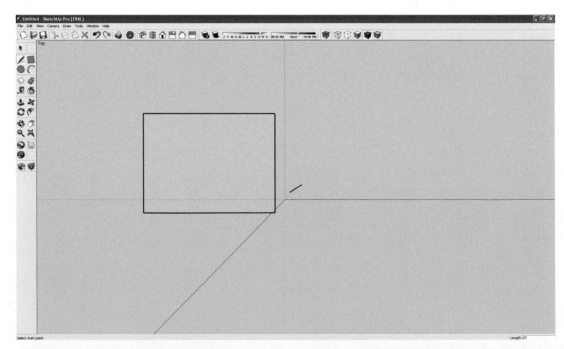

FIGURE 2–7 A series of closed lines creates a skin or surface, represented in purple.

Selection Methods

What you select will dictate which method for selection to use. Activate the Selection tool to select a line or surface of the box. Hit ESC to deselect. Double-click on an object to select all adjacent pieces of the box. To add different objects to a selection, hold down the Shift key. This allows you to both add to a selection and take selected items out of a selection. Later when we get into 3D, you can triple-click on an object and it will select all joined pieces.

Erasing

There are several methods for erasing. Activate the Pink Eraser tool and swipe it over the lines of the box. This will erase the line, which in turn will eliminate the skin that was created from closing the box (figure 2-8). Select the Undo arrow in the Standard toolbar to get your box back.

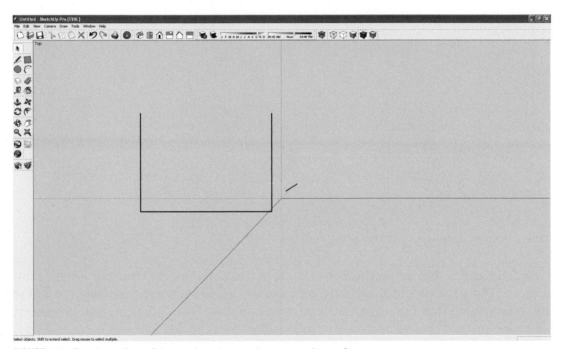

FIGURE 2–8   Erasing a line of the enclosed area also erases the surface.

You can also erase by activating the Selection tool, select a line, and hit the Delete button. This results in a three-sided shape with no surface similar to what happened when using the eraser. To erase the skin but not the lines, you can right-click on the surface and select Erase.

Undo

To undo the Erase command created above, you can select the Undo arrow on the Standard toolbar, go up to **Edit**—Undo, or hold down Control+Z.

Undo the eraser command so that you still have your rectangle with a surface.

Splitting a Surface or Line

When subsequent lines are drawn on an existing skin, the new line breaks the edge of an existing line and/or the surface of an existing skin. Use the Line tool to draw a rectangle in the lower right-hand corner of the rectangle (figure 2-9). Use the inference (on edge) to make sure you are touching the line. You will notice that the new lines are thin compared to the thicker outlines. This indicates that the surface has been cut and is now two separate surfaces. The right and bottom lines have also been cut and are now considered two line segments each.

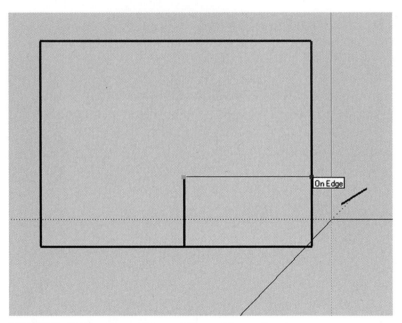

FIGURE 2–9   Cutting the surface and lines of the rectangle.

You can check this using the Selection tool and selecting the bottom edge of the rectangle. The selection tool will highlight the line edge up until it reaches the break point, recognizing that the edge has been cut by another line. Similarly, you can select the surface and see that each area is now considered a separate space.

Arc

The Arc tool is actually a series of short line segments that simulate the curvature of an arc. Activate the Arc tool, select the top right corner of the large rectangle, and then slide down the side to select On Edge of the line segment of the inside rectangle. This will be the end point of the arc. Then drag out the cursor to the right. At this point you can select a third point for the arc or type in a radius (figure 2-10). Type in a radius of 4', and then hit Enter. To get a smoother arc, type in 20s (20 segments) and hit Enter.

Notice that the arc outline is now the bold line and the line segment from the rectangle is a thin line. This demonstrates that the two surfaces are connected and that the arc is closed.

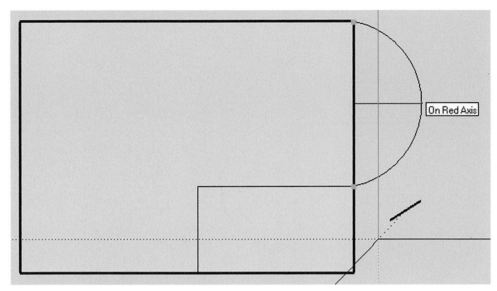

On Red Axis

FIGURE 2–10   Creating an arc.

Healing the Surface

Healing the surface refers to getting rid of those thin inside lines, such as between the arc and the rectangle, so that it reads as one entire surface. Use the pink eraser to rub out the thin line running between the arc and the main rectangle. That surface is now considered healed.

Circle and/or Polygon Tool

The Circle tool is very similar to arc in that it is composed of a series of 24 line segments and accepts information via the VCB. Activate the Circle tool and select a point to the right of the rectangle and start dragging out your cursor to see a circle form. Now you can enter in a radius or select a point. After the circle is drawn you can enter in a number of sides to smooth it out (figure 2-11). You can change either the radius or segments as many times as you wish as long as you do not start another command. To change the radius or number of line segments after another command has been activated, use the Selection tool and select on the edge of the circle, polygon, or arc, then right-click on the edge and choose Entity Info. Here you can change the number of sides it has or the radius.

Polygons are created by using the Circle tool and giving it fewer segments. Try creating an 18″ polygon with six sides.

Copy

Use the Selection tool, double-click the polygon you made, and activate the Move tool while holding down the Control key. Select the polygon again and drag it down along the green axis. Now you can select a point for it to copy to or release the Control key and type in a distance. Type in 3′4″ as the distance for the copy to move (figure 2-12).

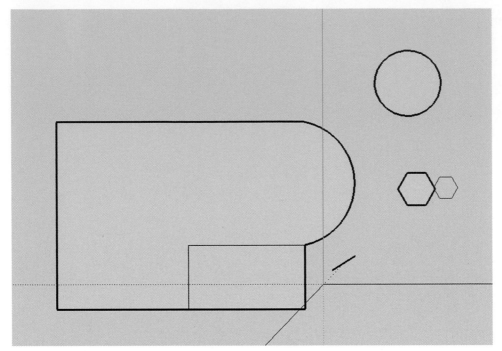

FIGURE 2–11 Circle tool with a different number of line segments.

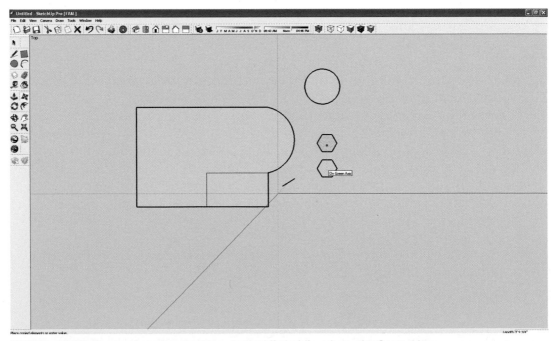

FIGURE 2–12 Making copies using the Move tool while holding down the Control key.

To make multiple copies, type in 3×(figure 2-13). This tells SketchUp to make a total of three copies using the original copy distance and direction as a guide.

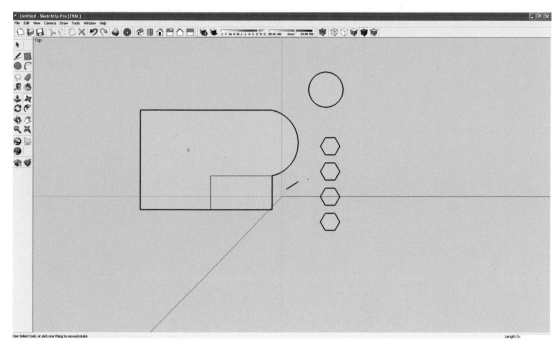

FIGURE 2–13 Multiple copies.

Rotate

The Rotate tool allows you to select items for rotation. Activate the Selection tool and window around the four polygons created earlier. Now activate the Rotate tool and select the center of the bottom polygon. The second selection point you make demonstrates the original orientation; drag your cursor and select a point anywhere along the green axis. The third selection point you make dictates the direction of the new orientation; drag your cursor to the left along the red axis and select a point anywhere along the red axis. Use the Selection + Shift key to select all the polygons and then the Move tool to move the polygons to the lower right corner of the rectangle (figure 2-14).

Rectangle

The Rectangle tool creates rectangles, which also receive information via the VCB. Activate the Rectangle tool, pick a point on the upper left of the screen, and start to drag your cursor. At this point you can select the opposite corner of the rectangle or enter in two distances separated by a comma. Typing in 1',1' will result in a 1 square foot box (figure 2-15).

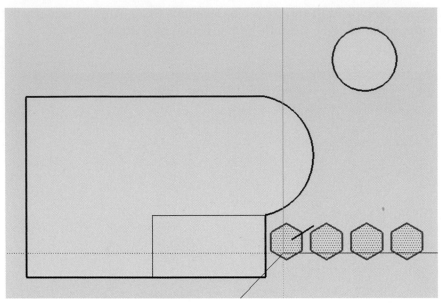

FIGURE 2–14 Using Rotate and Move.

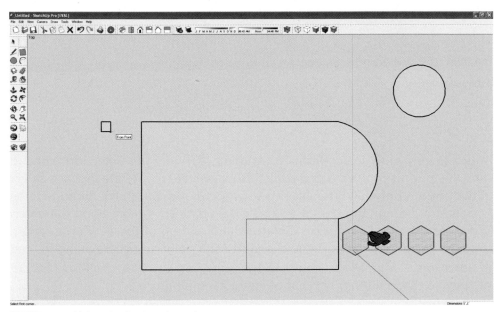

FIGURE 2–15 Using the Rectangle tool.

 ## Basic 3D Drawing—Let the fun Begin!

Orbit

Orbit is an interactive command allowing you to move more freely within your 3D model. There are two ways to use it. You can activate the Orbit tool, then click and drag in the upward direction. At this point you will see your blue axis. Or you can also hold down the roller bar of the mouse and drag it up (figure 2-16). To get back to plan view, click on the Top View tool from the **View** toolbar.

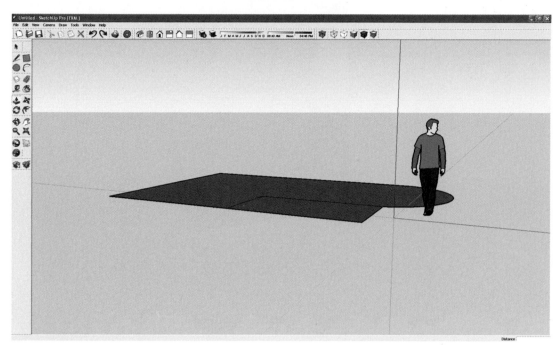

FIGURE 2–16 Orbit.

Use the Selection tool and window around the smaller shapes to delete them. Holding down the Shift key will allow you to also select the square.

Push/Pull Tool

The Push/Pull tool generates 3D modeling. Activate the Push/Pull tool, select the surface of the large area, and drag your cursor up along the blue axis. At this point you can visually select another point or you can type in a specific height. Type in 15′ and you will see the structure rise 15′ (figure 2-17).

Offset Tool

The Offset tool allows you to offset lines and surfaces at a specified distance. Activate the Offset tool. When you hover along the edge of the top of the structure, you will see a small red dot that runs along the edge as you move your cursor along the edge.

This small red dot picks up all the line segments in more intricate configurations that you wish to offset. For this simple structure, you do not need to run it along all the edges—just pick a point inside the surface area and start dragging it to the inside of the structure. At this point you can visually select another point or you can type in a specific distance. Type in 1′6″ and you will see the top edge offset to the inside 1′6″ (figure 2-18). You can now create a sunken roof by using Push/Pull to push down the inside portion of the roof (figure 2-19).

FIGURE 2–17 Using the Push/Pull tool.

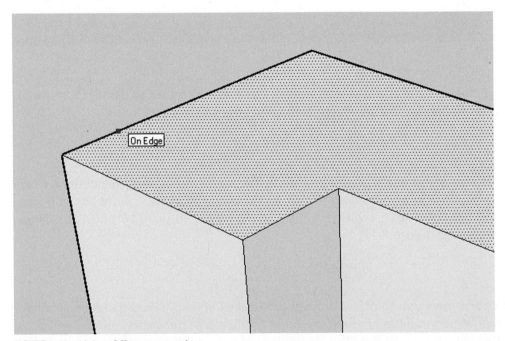

FIGURE 2–18 Using Offset on an edge.

Creating a Simple Structure

Start a new SketchUp drawing to work through this part of the tutorial.

Draw a 50′ × 30′ rectangle by activating the Rectangle tool; pick a point in the upper left side of the screen and start dragging it to the lower right of the screen and type 50′,30′.

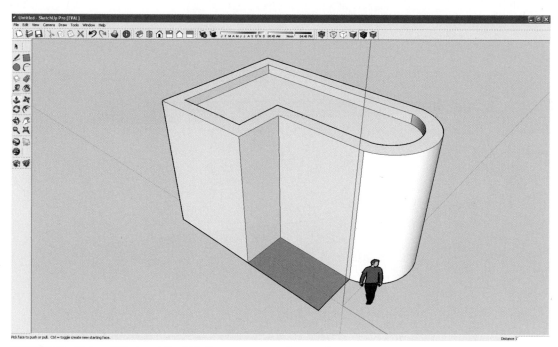

FIGURE 2–19 Using Offset and Push/Pull to create a sunken roof.

Use the Zoom Extents tool to zoom out to the entire drawing space. Orbit so you see the rectangle in perspective as seen in figure 2-20.

Use Push/Pull to raise the structure 12′ (figure 2-21).

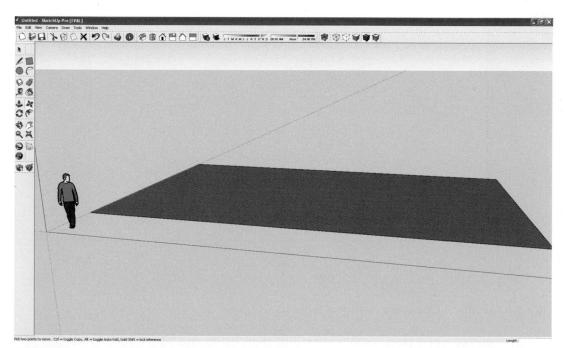

FIGURE 2–20 Using the Rectangle tool to create a surface.

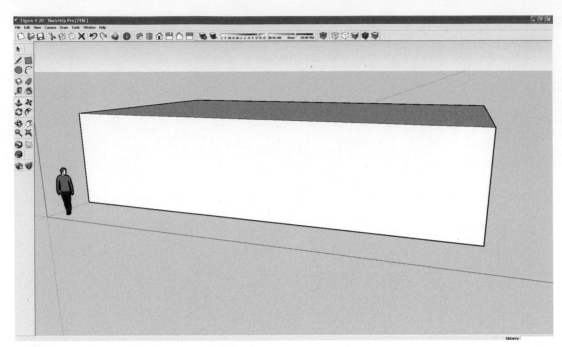

FIGURE 2–21 Using Push/Pull to create height.

Select the Top View and draw a line from midpoint to midpoint along the length to create a roof line (figure 2-22).

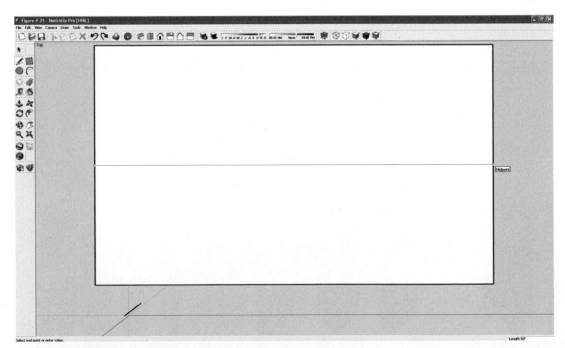

FIGURE 2–22 Plan view for drawing a center line for the roof.

Orbit again to get a perspective view and then using the Move tool select the line you just drew to pull the roof line up along the blue axis, then type in 8′ (figure 2-23).

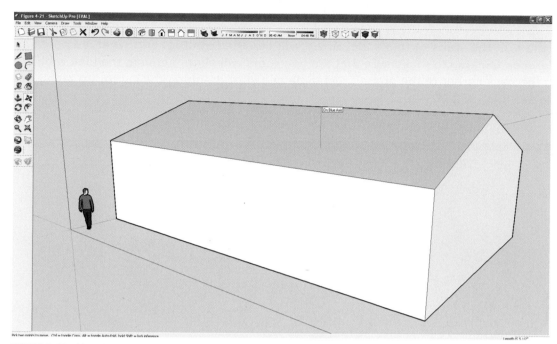

FIGURE 2–23 Use the Move tool to pull the center line up along the blue axis.

Now we will make a small addition to the side of the house. Draw a rectangle on face along the length of the house. The sample below uses a 10',12' rectangle. Draw another rectangle along the ground that matches up in length with the rectangle on the wall. The sample uses a rectangle that is 10',5' (figure 2-24).

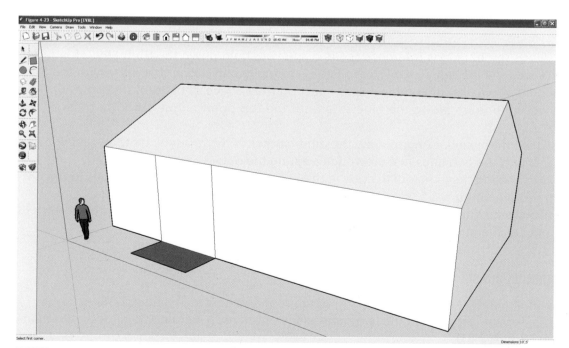

FIGURE 2–24 Use the Rectangle tool to draw on the footprint of the addition on the house and on the ground.

Inference Locking

To draw in the walls of this addition we will use the Line tool and what is known as **Inference Locking** to make sure that the two roofs will align. Activate the Line tool; select the back corner of the ground rectangle and start dragging the cursor up along the blue axis (figure 2-25).

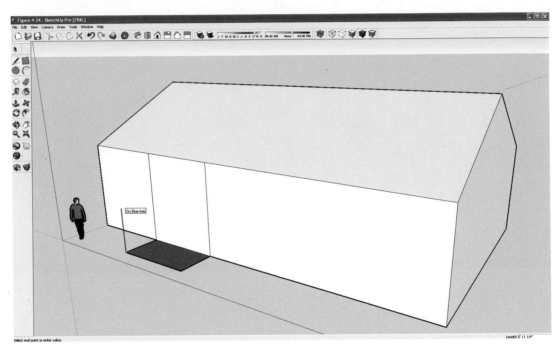

FIGURE 2–25   Using Inference Locking to draw.

Once the blue axis inference tag is displayed, hold down the Shift key. This refers to locking the inference and forces that line to move only along that axis. With the Shift key held down, locking on blue axis, select on the surface of the roof (figure 2-26). This draws the line up along the blue axis until the top point aligns with the pitch of the roof.

Use Inference Locking to draw the other corner of the ground rectangle. Draw a line connecting the two lines just drawn for creating the outside wall (figure 2-27).

Keep using the Line tool to draw in the lines from the corners of the walls to the bottom roof line (figure 2-28).

To create surfaces on the two sides, use the Line tool and draw lines from the top corners of the outside wall connecting to the roof line (figure 2-29).

Repeating a Task

Use the Push/Pull tool to pull out part of the roof and type in 1′ (figure 2-30).

To repeat this task on the two other sides of the roof, just double-click on their surfaces (figure 2-31).

Use the Line tool to draw in the remaining fascia (figure 2-32).

FIGURE 2–26 Hold the Shift key along the blue axis inference and hover over the roof.

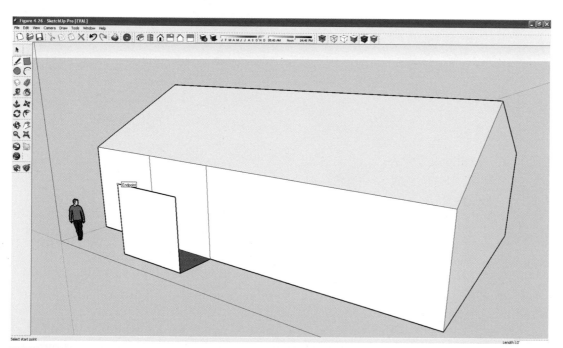

FIGURE 2–27 Use the Line tool to create a surface on the outside wall of the addition.

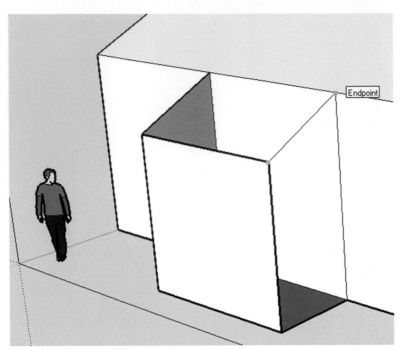

FIGURE 2–28   Use the Line tool to connect the corners of the surface to the roof.

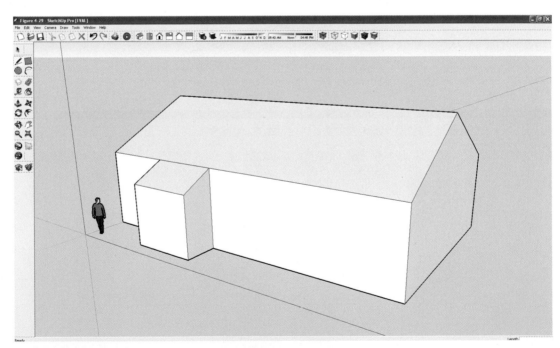

FIGURE 2–29   The addition should now have surfaces on every side.

Use Push/Pull to pull the fascia out 1′ and then double-click all the other surfaces to repeat this task (figure 2-33).

Zoom in to the peak of the roof and orbit so that you are looking slightly up at it. Using the Line tool, draw in the peak of the roof on one end using the inference tag parallel,

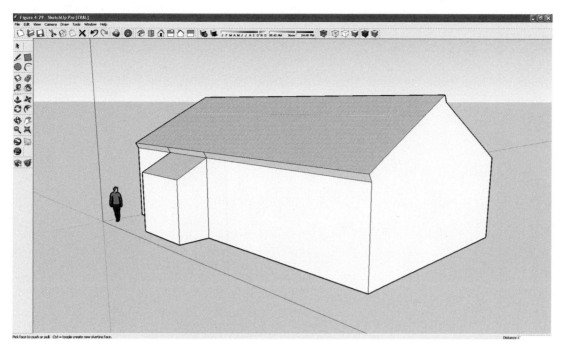

FIGURE 2–30 Push/Pull to pull out the roof.

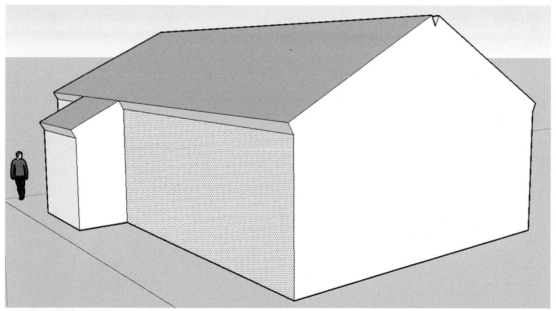

FIGURE 2–31 Double-click on the other surfaces to repeat the previous task.

which is represented by a magenta line, and the inference tag intersection created when hovering at the bottom of the peak (figure 2-34).

Orbit around to the other end of the house and use Push/Pull to drag the peak the length of the roof line (figure 2-35). Select a point along the far side of the roof line to get the on-face inference. This ensures that the peak runs flush with the other end.

Use the pink eraser to heal the surface of the fascia and roof (figure 2-36).

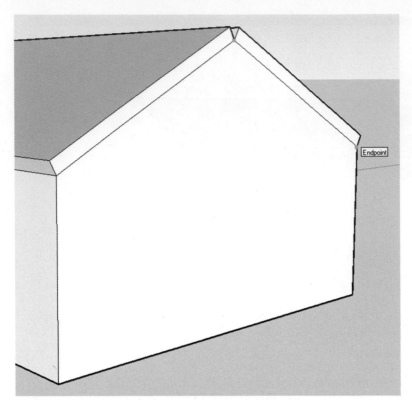

FIGURE 2–32 Use the Line tool to draw in the fascia.

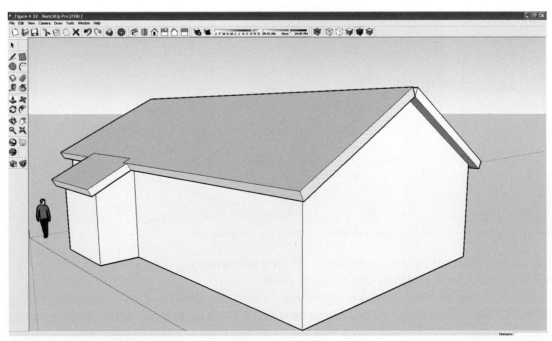

FIGURE 2–33 Use the Push/Pull tool to extend the fascia out 1′.

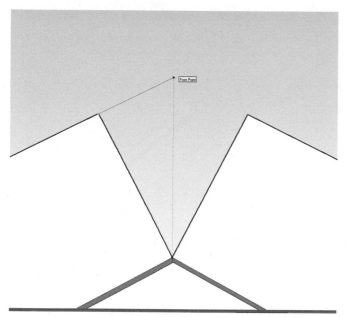

FIGURE 2–34   Drawing the roof peak using inferences.

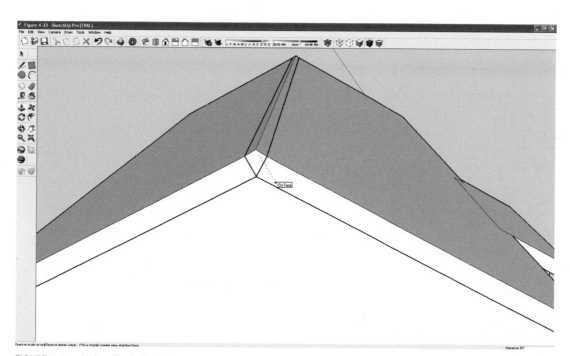

FIGURE 2–35   Using Push/Pull to pull the peak to the length of the roof.

Components

The components are found under Window — Components . You will see a variety of categories in the pull-down menu (figure 2-37). The components that appear are the samples. You can download more component files by going to http://sketchup.google.com.

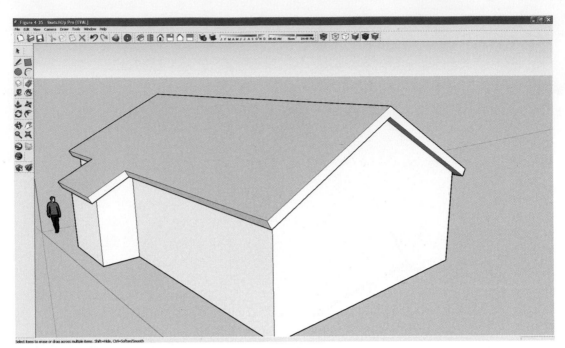

FIGURE 2–36 Roof lines are healed with the eraser.

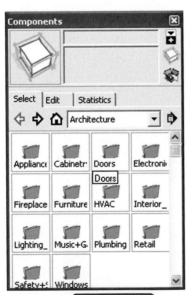

FIGURE 2–37 Components : Architecture components downloaded from SketchUp Google.

Select a download option and save it to your desktop. On your desktop, double-click the file to embed in the SketchUp program under program files.

Start by embellishing the structure with windows and doors. Select Architecture—Doors. Select a door and hover along the walls that you want your door to go on (figure 2-38). You should see the inference tag on face appear; now drag the door to the bottom of the structure so the inference tag reads On Edge, letting you know that the door is actually attached to the bottom façade of the structure.

In the Components palette, use the Back arrow to go back to the main listing of architectural elements (figure 2-39). Select the windows folder and add some windows (figure 2-40).

FIGURE 2–38 Adding doors.

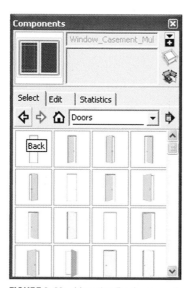

FIGURE 2–39 Use the Back arrow to get back to all of the Architectural components.

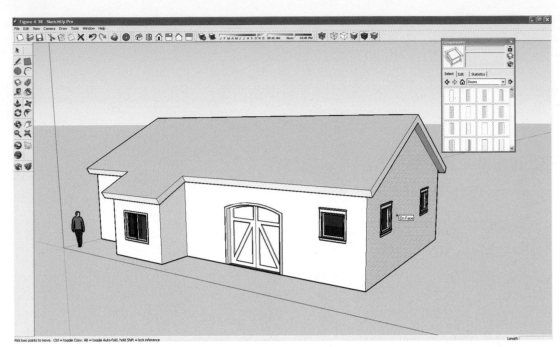

FIGURE 2–40 Adding windows.

> *author's note*
>
> When applying **Components** and **Materials**, the file size can get large very quickly. For this reason, the components and materials will be applied to two sides of this structure.

Face Me

In the **Components** palette, use the Back arrow again or pull-down menu to get back to the main subject headings and select the Landscape Architecture folder. Here you can find a variety of landscape features. Most components are 3D, but if you look under plant_materials, you will notice 2D and 3D options. The difference between the two, outside of the fact that one is 2D and the other is 3D, is that 3D images create a huge file size. The good news is that the 2D symbols have what SketchUp calls "Face Me" applied to them. This means that no matter which way you orbit around your model, the symbol will always face you. In plan view, they look like thin lines, but in perspective they will rotate around. I recommend putting in all the plant materials in plan view, especially if you are copying a plan already done in AutoCAD (figure 2-41).

You can add more items throughout your model. As you orbit around you might notice that some items get placed underground. If you are having trouble setting items on the ground, draw a small rectangle on the ground plain and then place the component on it. You can go back and erase the rectangle later.

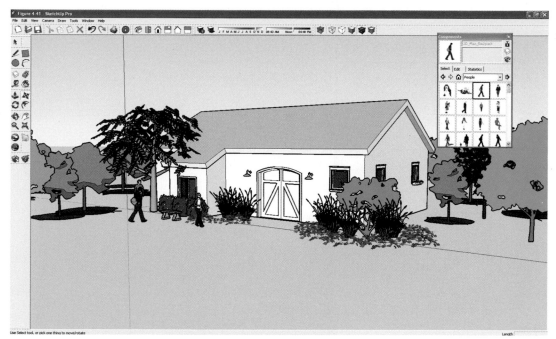

FIGURE 2–41 Adding landscape and people components.

author's note

At this point in the development of a project, many firms have enough information to print out several views and hand sketch accurate perspectives. Another option is to export this as a JPEG and render it in a more graphic-intensive program such as Photoshop or Piranesi. As previously mentioned there is some debate on whether the printed representational graphics of SketchUp are presentation quality. But below you will find a few extra steps toward finalizing the project for producing another popular method of project presentation: the slideshow and/or an animated walk-through.

Materials

The materials are found under **Window** — **Materials**. In the pull-down menu you will find a variety of materials that can be applied to your model (figure 2-42). Select Brick and choose the Brick_Rough_Dark, and now select anywhere on the surface of the building. To change it, select a different material and reselect the surface. If you want it to go back to the default colors, select the "set the material paint with to default," which is the purple/beige swatch on the material browser.

Altering Material

To exaggerate the scale of a material like the brick, select a material and then click on Create and alter the size or color of this material (figure 2-43). When you hit OK, it will be added to the materials list that exists in this model. Notice the tab in the

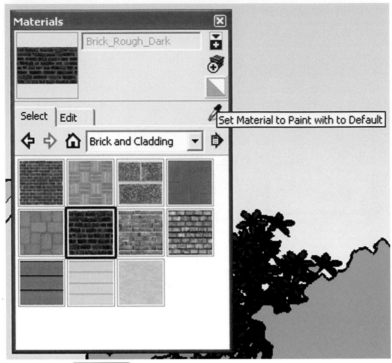

FIGURE 2–42 Materials.

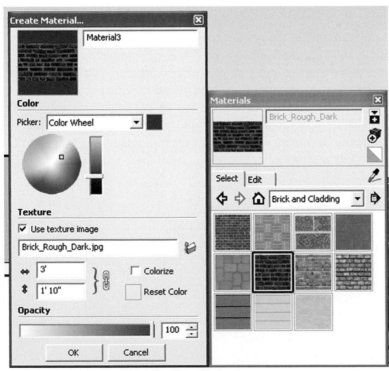

FIGURE 2–43 Creating new materials.

Materials pull-down menu has changed from **Materials** to In Model. Select the material here and then select the surface you want it applied to (figure 2-44).

FIGURE 2–44 Altered material applied to surfaces.

Setting Up a Slideshow

This feature sets up a series of pages that will run like a slideshow. The transition between pages is animated so it automatically continues to the next slide. To start, set the drawing to plan view by selecting the Top View from the **View** tools and selecting Zoom Extents. Then go to **View** — **Animation** — **Add Scene** (figure 2-45).

You will see a page tab appear at the top left of the workspace screen. Now orbit around and zoom in to your project and add another scene. You can change settings and effects when you add scenes, and those changes will be reflected in the slideshow. To hide components that will not be seen on a page, select the component and then go to **Edit** — **Hide**. This helps with running a smoother slideshow by allowing you to visually disable parts of your drawing that will not be seen. To turn them back on, go to **Edit** — **Unhide** — All or Last, depending on what you want to show up again.

Face Style

Face Style tools allow you to choose a variety of options that relate to how the model will be seen (figures 2-46 and 2-47).

When X-Ray is selected with any of the other options, it allows a see-through appearance of the model.

FIGURE 2–45 Setting up animation scenes.

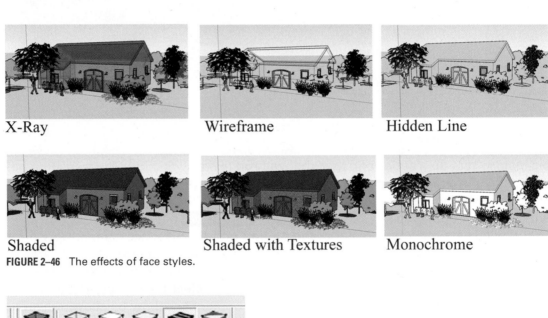

X-Ray   Wireframe   Hidden Line

Shaded   Shaded with Textures   Monochrome

FIGURE 2–46 The effects of face styles.

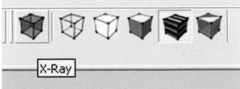

FIGURE 2–47 Face Styles: X-Ray, Wire frame, Hidden line, Shaded, Textures, and Monochrome.

Style Settings

The ⬚Styles⬚ settings allows you to choose a variety of options that relate to how lines will be seen. Go to ⬚Window⬚ — ⬚Styles⬚ and select the Sketchy Edges style and select each of the options to see how they affect the line work in the model (figure 2-48).

FIGURE 2–48   Different styles applied to the model.

You can also adjust the settings for many of the styles to exaggerate or minimize the effect by selecting a style and then the edit or the mix tab.

Add a slideshow page after selecting some display settings.

With the textures back on, it would look as shown in figure 2-49. Orbit around your drawing and add another slideshow page.

Shadow Settings

You can create quick shadow studies by using the **Shadow Settings**. Go to **Window** — **Shadows**. Here you can check the display shadows box and set a time of day, day of the month, and how light or dark you want your shadows to appear (figure 2-50).

FIGURE 2–49   Default style with textures applied.

FIGURE 2–50 Casting shadows using the **Shadow Settings** option.

But this shadow assumes north is to the right along the red axis and that it is in Boulder, Colorado. To change to a specific geographic location with a specific north angle, go to **Window** — **Model Info** and select **Location** on the left-hand list. You can choose a city, state, and north angle. In this example, I will set the shadows to be for Athens, Georgia, and north to be 45° from the red axis (figure 2-51). Add another page to the slideshow after setting shadows.

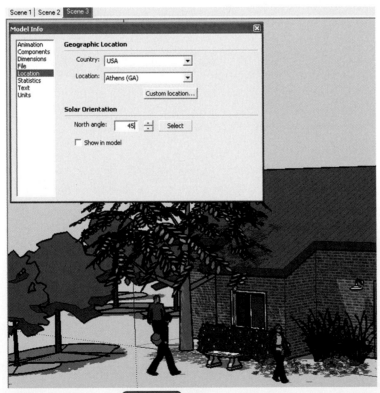

FIGURE 2–51 Choosing **Location** and north angle for shadow settings.

Text

To set a text style, go to [Window] — [Model Info] — [Text] from the list. Select the Font button to choose a font, a style, and a size. Hit OK to get back to the [Model Info] — [Text] and select the black box next to Font to change the color of the font. Hit OK to get back to the [Model Info] box. To type on screen, go to [Tools] — [Text] and select a location for the text to go. You can use the Selection tool to erase it, or double-click on it to edit the text. To change any existing text, make the changes in the [Model Info] — [Text] —Font button, make your changes, click on OK, and then choose the select all screen text from the bottom of the [Model Info] — [Text].

Slideshow Settings

Go ahead and add another slideshow page. At this point you should have several pages.

To set the timing for transitions go to [Window] — [Model Info] — [Animation]. This allows you to set timing for the transition between slides as well as the pause at each slide. Set page transition to 5 and page delay to 0. Right-click on one of the page tabs and select Scene Manager. Here you can change the name, organize your pages, or set properties for individual slides. Hit OK when done. Right-click on one of the page tabs again and select Play Animation to view the slideshow.

Saving and Animation

Save this as a SketchUp file by going to [File] — [Save As]. To save this as animation that runs smoother, go to [File] — [Export] — [Animation] and make sure the file type is set to .avi, which will play in the Windows Media Player. Once it has saved, open Windows Media Player and go [File] — [Open] and search for the file, select it, and open. Then select the play button at the bottom of the screen.

Saving and Printing

Save the drawing with the default SketchUp extension .skp to continue work on the model. You can also export the file as a 2D graphic with the .pdf, .jpeg, .tiff, or .dwg extension to be able to bring it into other programs.

To print, zoom and pan into the area you want to print. Go to [File] — [Print Setup] to select a printer, paper size, and orientation. This project is set to print to a color printer, ledger paper, and landscape orientation. Once the printer and paper are set you can go to [File] — [Print] and set it to fit to page.

If you want to print to scale, you must first deselect Perspective by exiting from the print dialog box and going to [Camera] — [Select Parallel Projection]. Then choose one of the predetermined views from the [View] tools. In the print dialog box, uncheck Fit to Page to have the option to select a scale by entering a unit in the printout to a unit on the ground in SketchUp. So a scale of 1/8″ would be a printout equal to 1″ and SketchUp equal to 8′.

Working through Lembi Park

Importing an Existing AutoCAD Drawing

SketchUp can import a variety of program files. The version of SketchUp you have dictates which version of other program files you can import. For instance, currently SketchUp 6 can only import AutoCAD 2004 documents or lower versions. With this in mind, you may need to save your AutoCAD file with a lower version before importing it into SketchUp. I also turned off all extraneous information. My first goal upon getting the project into SketchUp is to create a skin on all areas that require either texture or height. Extra information like the plants and hatch patterns and text will only interfere with making good clean surfaces in SketchUp. You can use the Lembi Park AutoCAD base generated from Chapter 1 or another AutoCAD plan for the following steps.

In SketchUp go to **File** — **Import** — **2D Graphic** (if you have the option). Make sure Files of Type is set to ACAD files; locate the file, open it and then close the Import Results box (figure 2-52). Select Zoom Extents to see your drawing.

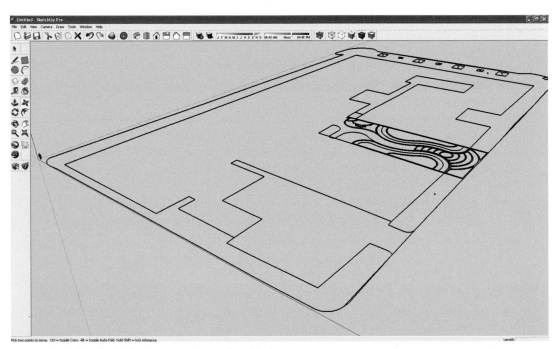

FIGURE 2–52 AutoCAD plan imported as a .dwg file.

Measuring Tape Tool

The Measuring Tape tool allows you to measure distances and scale items. To check the size of the drawing, select the Measuring Tape tool and select two points separated by a known distance (figure 2-53). For instance, the doorway at the lower left corner of the park area is 6'. Use the Measuring Tape tool across the doorway and the VCB should read 6'. If it does not, go ahead and type in 6' and hit Enter. A dialog box will ask if you want to resize the entire model: select yes so it scales everything up (or down) proportionately.

FIGURE 2–53 Using the Measuring Tape tool to check distances and/or resize the model.

Tracing

Tracing the project to create distinctive surfaces is a time-consuming and frustrating process, but it is necessary for showing textures and elevation. Use the Line tool, Rectangle tool, and Arc tool to trace over a majority of the surfaces. SketchUp has the ability at times to read existing AutoCAD line work and create a surface for that enclosed area. It may not always be the area you expected, but in most cases it will be an area that needs a surface. Go back and heal adjacent surfaces with the eraser (figures 2-54, 2-55, and 2-56).

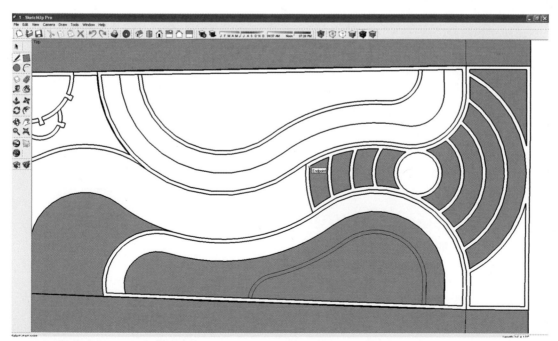

FIGURE 2–54 Tracing . . . little by little.

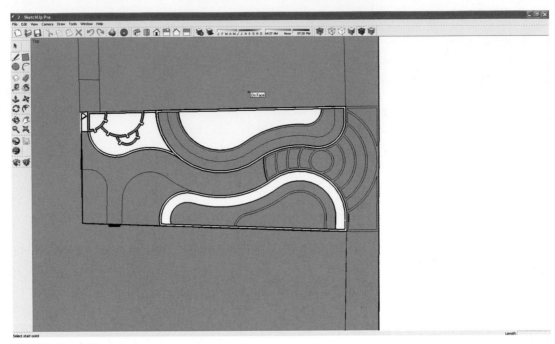

FIGURE 2–55 Still tracing.

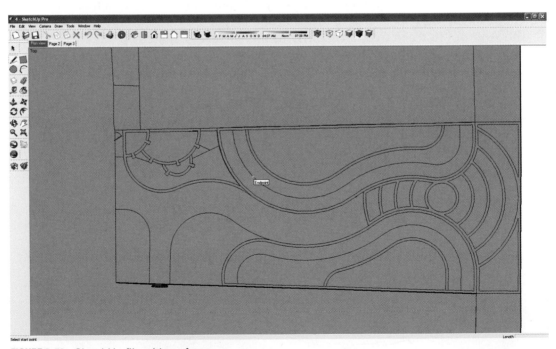

FIGURE 2–56 SketchUp file with surfaces.

Freehand Tool

You can trace areas using the Freehand tool by going to **Draw** — **Freehand Tool** and then selecting a line to trace by holding down the mouse button to draw along the edge of an existing line. The line should follow along the existing AutoCAD lines showing inferences along the way. When you loop back to the beginning, it should close and create a surface.

Checking Your Work

It is important that the lines you trace create separate surfaces. To check this, activate either the Push/Pull tool or Move tool and hover over surfaces. If more than one area is highlighted, then the surfaces are not split. Do not forget to use the Move tool to connect points.

Use the Push/Pull tool to bring the seat walls, planters, and water feature to the correct height (figure 2-57). It may help to Push/Pull the highest edges first and then the area inside of them. Start with the edge of the 4' planter and Push/Pull it 4' up along the blue axis. Then Push/Pull the area inside of it to 3'8", creating a small lip.

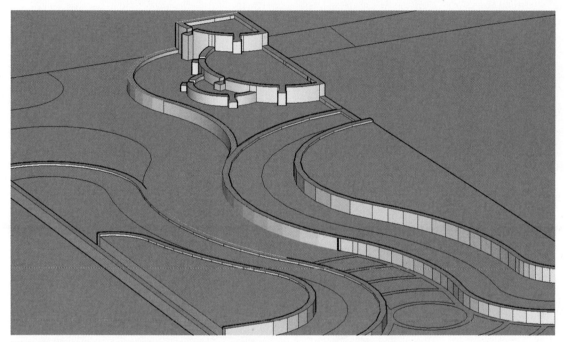

FIGURE 2–57 Using Push/Pull to elevate the planters, seat wall, and water feature.

Set the view to X-Ray so you can see through the elevated seat and planter walls (figure 2-58).

Layers

The layers created in AutoCAD are automatically brought over with the imported document. Go to Window — Layers (figure 2-59). A simple layers palate with a radio button indicates which layer you are currently on, the layer names, their visibility, and their color. You can now turn the plants back on by checking the layers with plant names. The following layers were turned back on to locate plant materials:

QV thick

AV5 heavy

Shrub heavy

Shrub A heavy

Shrub B heavy

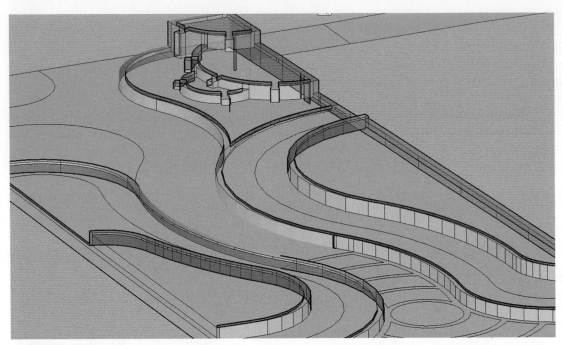

FIGURE 2–58 X-Ray view.

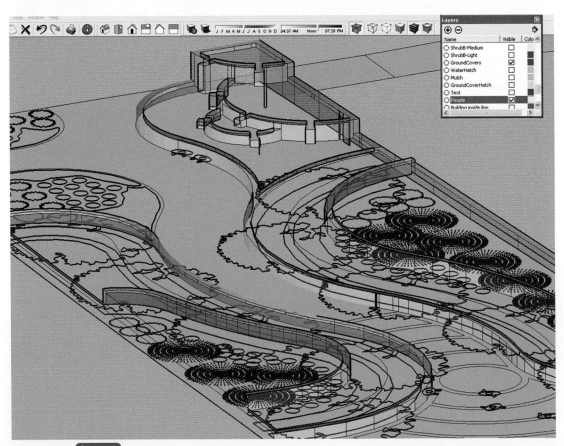

FIGURE 2–59 Layers turned on.

Shrub C heavy

Ground Covers

People

With the X-Ray view on, you will be able to see all the plant symbols. As you place the plant materials, make sure the inference reads on face, which should place your plant material on the surface of the elevated area (figure 2-60).

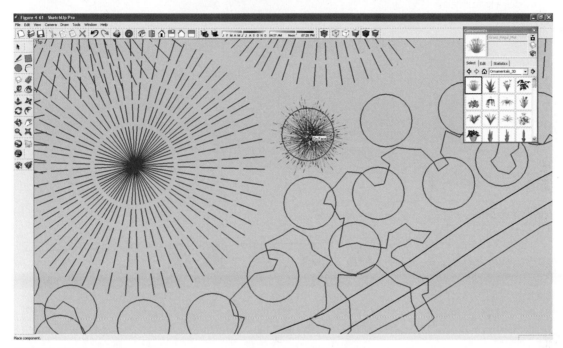

FIGURE 2–60 Place components on face so they do not go above/below ground.

Continue placing components and materials. For a better visual effect and to keep the file size smaller, only place plant materials that create the scene along the route you will be viewing and leave the others off (figure 2-61). Also consider working in shaded view and with 2D components (figure 2-62). Once a group of plant materials have been placed, you can go back and turn the layers off so it gets rid of (visually turns off) the AutoCAD line work. This will also help with file size. Elevate the surrounding buildings for a sense of enclosure, and remember to save your drawing often.

Group, Hide, and Make Component

Grouping like items or making them into components reduces regeneration time, allowing the computer to work faster. Select the items you want to group or make into a component, then go to **Edit**—Make Component or Make Group. To edit the group or component once it is made, go back to **Edit**—Group and select from the options available:

Edit Group—to edit entities inside of a group such as adding material texture or color

Explode—to undo the group

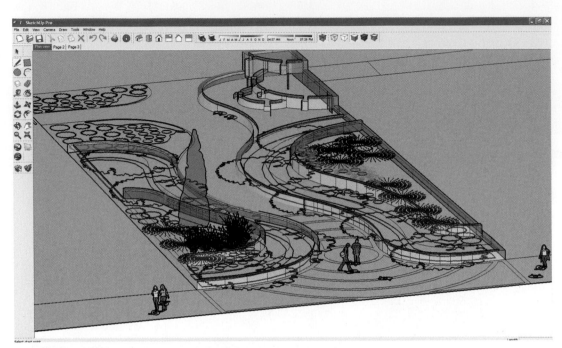

FIGURE 2–61 Placing components and checking views.

FIGURE 2–62 Using styles to check views, materials, and component accuracy.

Make Component—to turn your group into a component

Hiding items allows for faster regenerations and also allows easier viewing of tight areas. Select the items you want to hide and go to **Edit** — **Hide** . To turn them back on go to **Edit** — **Unhide** —All or Last (figure 2-63).

FIGURE 2–63   Similar trees were grouped and then hidden.

In preparation for finishing this project, I have selected a view which best depicts a specific shot I'd like to use for printing (figures 2-64 and 2-65). To keep consistent with this view, I created a scene tab for it that allows me to go back and edit exactly what will be seen in the printout. You might consider creating two to three scenes just in case the development of one does not go as planned or print as well.

FIGURE 2–64   Finished view of the Park project.

FIGURE 2–65 The same view with different effects applied to it: a watercolor version and a sketch version.

TERMS

Axes—SketchUp uses a 3D coordinate system, similar to AutoCAD, where points in space are identified by position along the drawing axes identifiable by their color: red, green, or blue.

Blue Axis—The blue axis is represented by a solid blue line (for positive numbers or items above ground) or dashed blue line (for negative numbers or items below ground). This refers to height or the Z axis in AutoCAD.

Green Axis—The green axis is represented by a solid green line (for positive numbers) or dashed green line (for negative numbers). This refers to depth or the Y axis in AutoCAD.

Healing—This refers to erasing the thin inside lines that have broken a surface, so that it reads as one entire surface.

Inference Locking—If you want to maintain an inference, like staying on axis, while drawing or moving, hold down the Shift key. This will lock the inference so that, regardless of where you move the cursor, the item you are drawing or moving stays on the selected inference.

Inference Tags—These are small interactive tags that appear at the end of your cursor, which locate axis, points, lines, and surfaces.

Red Axis—The red axis is represented by a solid red line (for positive numbers) or dashed red line (for negative numbers). This refers to width or the X axis in AutoCAD.

Skin or **Surface**—When a series of line segments are connected and closed or when a shape, such as a circle, is drawn, the inside area receives a surface.

Value Control Box or **VCB**—This area located within the status bar receives typed information and also reads dimensions of items being drawn.

from AutoCAD to Adobe Photoshop CS2 Rendering

By Professor Jose R. Buitrago

CHAPTER 3

CHAPTER OBJECTIVE

This chapter introduces readers to the step-by-step process of exchanging (importing) an AutoCAD drawing into **Adobe Photoshop** CS2. This program interchange will allow readers to learn how to render an AutoCAD document using Adobe Photoshop CS2. By the end of this chapter you should be able to render your CAD drawing using several rendering techniques in Photoshop such as air brush, blue print, sepia, water color, color pencil, and color marker finishes.

Introduction

There has always been a general unwritten consensus among landscape architects that computer-aided design (CAD) drawings are too sterile to showcase the best attributes of any design. CAD drawings have also been described as cold, generic, and bare. In sharp contrast, traditional hand-drawn media is considered the absolute best format to illustrate "good design," and the true measure for gauging the talent of the designer. The general assumption is that CAD is solely used for assisting in the creation of construction documents and not to be used for other "creative" uses. In the early years of the development of the CAD software, most of these assumptions were certainly true, but recent advances in computer technology are challenging these views.

The newly revised and improved versions of CAD have begun to incorporate "color rendering" tools such as **M-Color** (a plug-in for CAD). Although these improvements are directed toward the needs of the CAD–engineer base market, they are still in the early stages of development. On the other hand, CAD is becoming more flexible and compatible with other rendering software such as Adobe Photoshop CS2. This compatibility now offers a full spectrum of color rendering formats available only in Photoshop CS2, which surpasses the current version of M-Color CAD plug-in. This chapter will guide readers in a step-by-step process that shows how to integrate CAD and Photoshop.

First Step: Importing a CAD Drawing into Adobe Photoshop

The first step is to launch the latest version of AutoCAD and open a drawing that has clear, precise, and strong graphics. Your CAD drawing must be in paper space at the scale to be plotted. Please refer to the AutoCAD (Chapter 1); Setting Up the Drawing (figure 3-1).

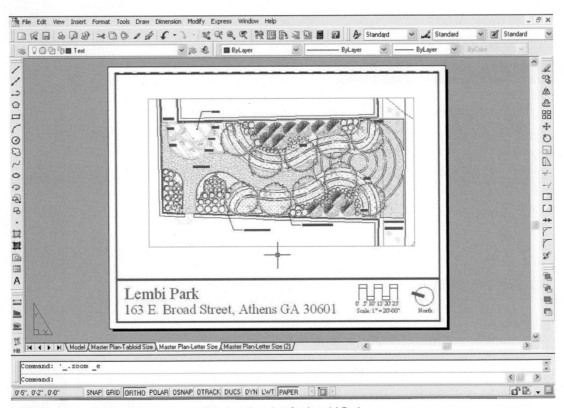

FIGURE 3–1 The Author's CAD screen showing the plan for Lembi Park.

Second, under the AutoCAD top menu, `File`, select `Plot File`, and change the plotter to **Adobe** PDF. Confirm that your current drawing page layout matches the plot settings by clicking on the Properties button of the `Plot File` pop-up menu window. This will be followed by another pop-up menu window; now click on `Custom Properties` (figure 3-2).

On the `Custom Properties` window, check that the `Adobe PDF Settings` —Default is set for high-quality print, the `Paper/Quality` is selected to be in the correct Paper Source—Size, the Color is selected as Black and White, and the `Layout` is selected according to the drawing specifications (figures 3-3, 3-4, and 3-5).

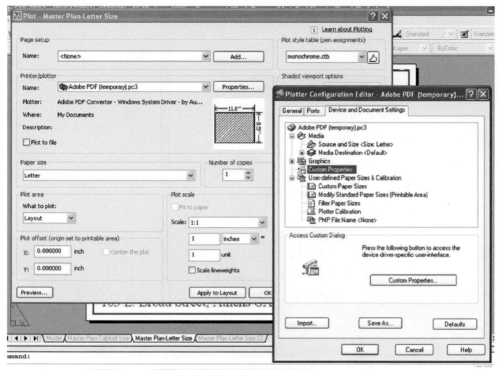

FIGURE 3–2 The Plot and Plotter Configuration Editor pop-up menu window.

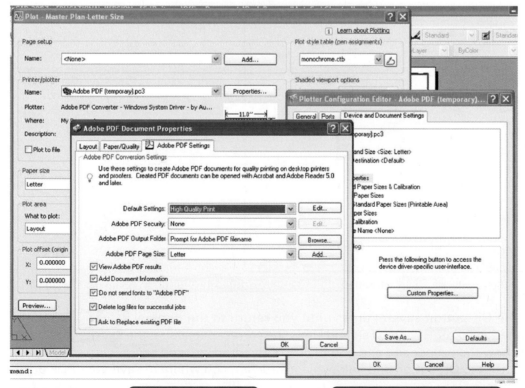

FIGURE 3–3 Setting the Adobe PDF Settings tab under the Plotter Configuration Editor pop-up menu window.

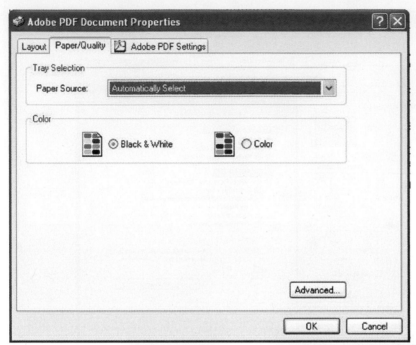

FIGURE 3–4 Setting the `Paper/Quality` tab of the `Adobe PDF Document Properties` pop-up menu window.

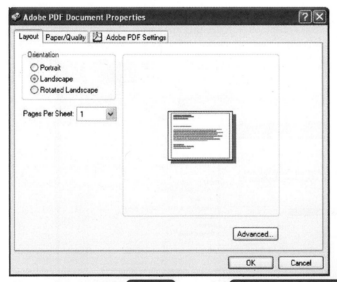

FIGURE 3–5 Setting the `Layout` tab of the `Adobe PDF Document Properties` pop-up menu window.

Hit the OK button several times until you return to the original `Plot File` window. There, the `Paper Size` window should show the correct paper size, the `Preview` window should show the proper size and orientation of the drawing, and `Plot Scale` should be a 1:1 ratio. Hit the OK button and follow the instructions on where to save your PDF document (figure 3-6).

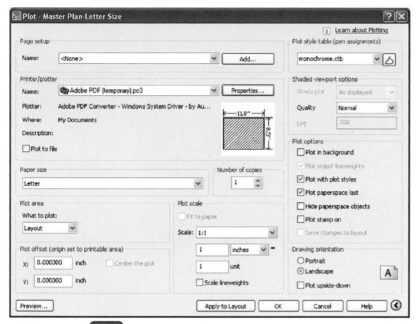

FIGURE 3–6   The **Plot** pop-up menu window.

author's note

Make sure that the paper size and orientation match the properties of your original CAD drawing. If this does not work, then select under **Adobe PDF Document Properties** — **Layout** — Advanced, the "Postscript Custom Page Size Definitions" instead of selecting from **Paper Size** drop-down menu (i.e., ARCH A, ARCH B, ARCH C, ARCH D, ARCH E, ANSI A, ANSI B, ANSI C, ANSI D, ANSI E, and so on). A new window will pop up and you will be required to type the correct paper size according to the paper feed direction (longest side first). Hit OK when you are done. This will return you to the previous window. Continue selecting OK, **Save**, or hit Apply button until you return to the first **Plot** window, and then hit once more OK to create the PDF CAD file. It is important that the **Preview** window of the **Plot Size** window shows the correct layout orientation and paper size.

Your browser might automatically launch **Adobe Acrobat Professional** and open the document. If you follow these instructions and get to this stage, congratulations, you are ready for the next step (figure 3-7).

Close AutoCAD and Adobe Acrobat and launch Adobe Photoshop CS2. Under **File** menu, select **Open** and browse until you find your Adobe PDF drawing, then hit the OK button (figure 3-8). The **Import PDF** pop-up menu window will open. There, on **Select**, choose Page, Thumbnail—Size—Fit the Page, Crop—Bounding Box, Resolution—300 dpi or larger, Mode—RGB, and Bit Depth—16 bit. Hit OK (figure 3-9). Your drawing should carry the same paper size dimensions of your CAD drawing (figure 3-10).

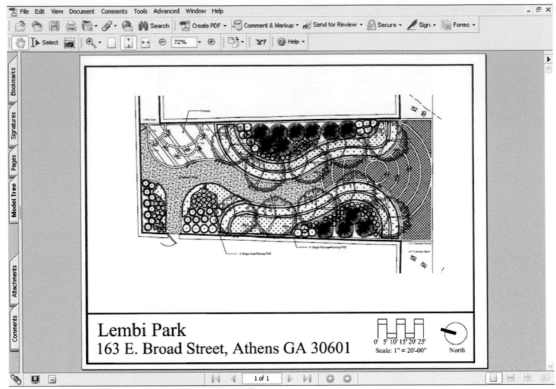

FIGURE 3–7 The ◖**Adobe Acrobat Professional**◗ window preview of the CAD drawing save as a PDF format.

FIGURE 3–8 The Adobe Photoshop CS2 ◖**Open**◗/Browse selection pop-up menu window.

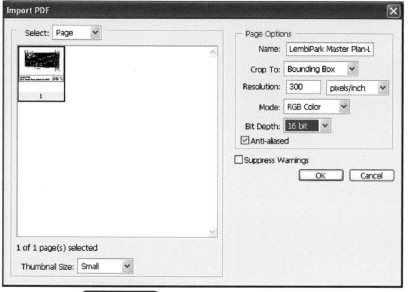

FIGURE 3–9 The **Import PDF** pop-up menu window showing the page options.

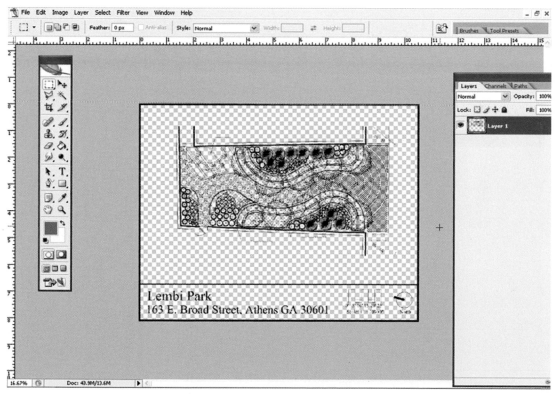

FIGURE 3–10 The Adobe Photoshop CS2 window screen showing the imported Lembi Park PDF file image.

If not, just under the top menu option **Image**, select **Canvas Size** and match the canvas size to the size of your original drawing (figures 3-11 and 3-12). Be aware that there is always a small percentage of distortion in changing the AutoCAD file into a PDF and then into Photoshop CS2. This is normal but should not be noticeable to the trained eye (figure 3-13).

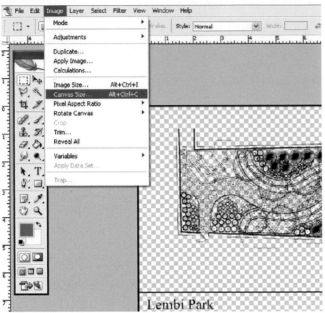

FIGURE 3–11 Under the **Image** top tab, scroll down, and select **Canvas Size**.

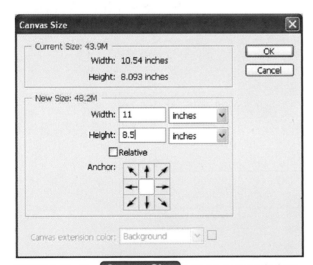

FIGURE 3–12 The **Canvas Size** pop-up menu window.

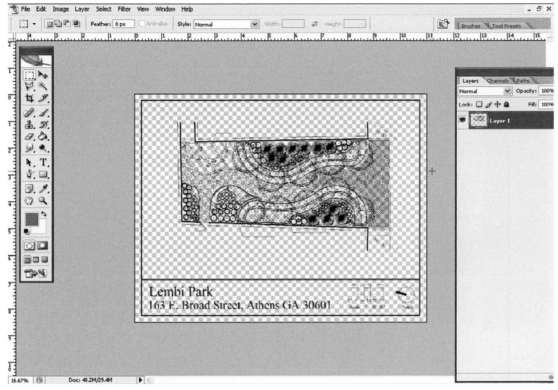

FIGURE 3–13 The Adobe Photoshop CS2 standard screen.

At this point you have learned how to import a CAD drawing into Adobe Photoshop CS2. The imported PDF drawing should be in the layer menu as standard Layer 1. Depending on your Photoshop default setting, your background should be either white or transparent. Now you are able to make full use of the different rendering techniques available on Adobe Photoshop CS2 such as blue print, sepia print, water color, color pencil, color markers, air brush, and many others. Program Interchange and Student Project Examples (Chapter 6) of this book will provide further instructions and illustrations of these rendering techniques. The next step will focus on the popular air brush color rendering technique.

Air Brush—Color Rendering the PDF CAD File Using Adobe Photoshop CS2

In 1879, Abner Peeler of Iowa invented an air-operated tool that sprayed dye, ink, and water-base paint. This tool was quickly adopted by graphic designers and illustrators because of the fast, economical, and easy application of color to large drawing areas. Architects and landscape architects adopted this tool in their rendering palette because of the soft and transparent character of this media. Today, because of the purchase cost of basic air brush nozzles and air pump kits, this format has lost its advantage over other formats such as color markers and pencils. This format can be easily recreated using Adobe Photoshop CS2 by following these simple steps.

First, under the [Layers] tab, create a new layer and label it Background. Rename the PDF drawing layer as Line Drawing. This simple step allows us to quickly recognize where drawings are located (figure 3-14). Next, make the Background layer the active layer and turn off the Line Drawing layer. On the toolbar, click on "Set Background Color." This should launch the [Color Picker] window. Scroll up or down on the color bar to get the white schemes tones and value. Select white as the background color, and then hit OK (figure 3-15). RGB values of white are R/255, G/255, and B/255, or Number ffffff.

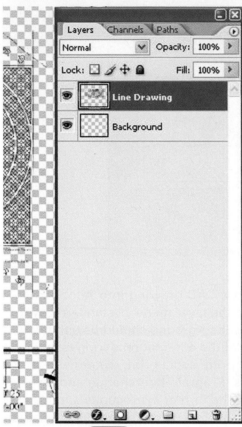

FIGURE 3–14   The [Layer] floating window tool.

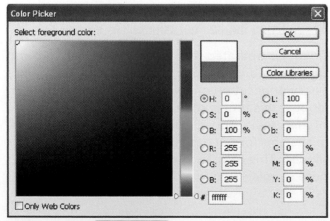

FIGURE 3–15   The [Color Picker] pop-up menu window.

Selecting white as the background color will provide higher contrast for color rendering. Using the Rectangular Marquee tool, select the entire Background layer. Select the Bucket tool and select the white color from the background color, and click inside the field previously selected with the Rectangular Marquee tool. Make sure the Background layer is active. This action will fill the Background layer with a solid tone of white color. After completion, deselect the area (figure 3-16). Remember that the Background layer is under the Line Drawing layer, so make sure that your layer order is correct; otherwise, the line drawing might end up underneath the solid field of white.

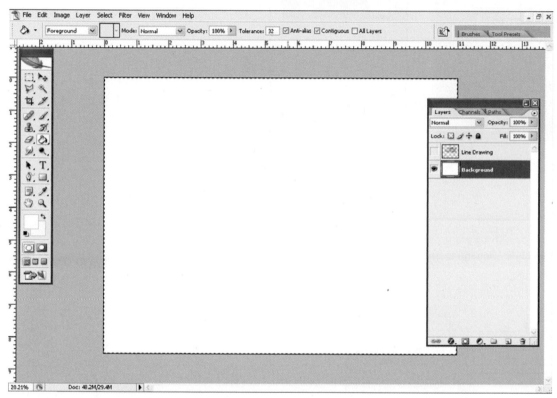

FIGURE 3–16 The Background layer color render white.

Now turn the Line Drawing layer on and make it the active layer (figure 3-17). The next step is to create a new layer, between the Line Drawing and Background layer. This is done in order to prevent the black line drawing being covered by the color layer. Remember that the Layer tabs reads from top to bottom, and thus it will follow the same order of vertical placement on the drawing. Label this new layer Color Render 1. Make the Color Rendering layer the active layer (figure 3-18).

The next step is to select the Brush tool that most closely resembles the air-brush technique. In the Adobe Photoshop CS2 floating toolbar, select the "Brush Tool (B)" icon. This will launch, under the top toolbar, the Brush Size, Mode, Opacity, Flow, and Set to Enable Airbrush Capabilities icons (figure 3-19).

FIGURE 3–17   Renaming the layers at the Layer floating tool window.

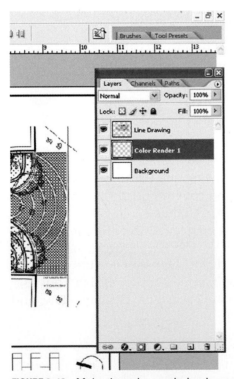

FIGURE 3–18   Make the color rendering layer the active layer.

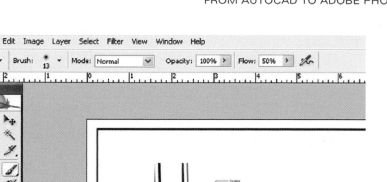

FIGURE 3–19   In the floating toolbar, select the "Brush Tool (B)" icon.

Next to the Brush Icon of the top toolbar there is an arrow. Click on the arrow and this will launch a pop-up menu window. In the top of this window there is another arrow. Click on this arrow to launch a second pop-up menu with side scroll bar. Scroll down the list and select **Basic Brushes** (figure 3-20). A pop-up window will appear, asking you to accept changing your current Brush Style options to **Basic Brushes**. Hit OK to accept making **Basic Brushes** the active brush selection.

The first **Brush Option** pop-up menu window now shows your current Basic Brush options. Use the scroll up/down side bar to locate and select any of the Soft Mechanical brushes. The Soft Mechanical brushes vary in size from 1 pixel to 500 pixels. Start by selecting the Soft Mechanical 100 Pixel as a basic. Set the Master Diameter to 100 pixels, and the Hardness to 0% (figure 3-21).

Once the Soft Mechanical brush is set, the next icons under the top of the toolbar should be set as follows: the Mode set to Normal, the Opacity to 50%, the Flow to 50%, and Set to Enable Airbrush Capabilities should be turned ON. This will allow you to create the transparent and soft look created with air brush (figure 3-22). The next step is to select your color choices from the **Color Picker** box by clicking first on the Set Foreground Color in the floating toolbar (figures 3-23 and 3-24).

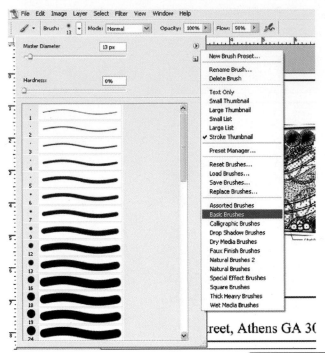

FIGURE 3–20 Scroll down the list and select the `Basic Brushes`.

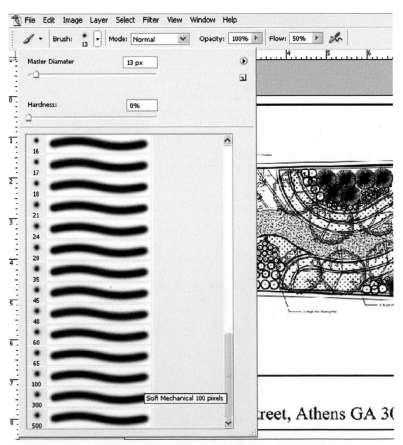

FIGURE 3–21 Set the Master Diameter to 100 pixels, and the Hardness to 0%.

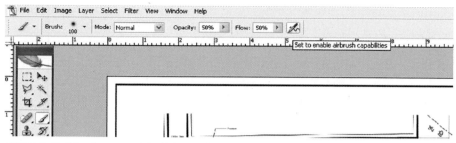

FIGURE 3–22   The Set to enable Airbrush Capabilities should be turn ON.

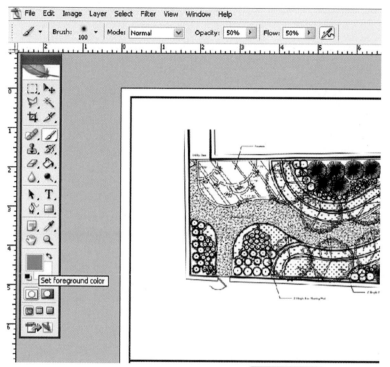

FIGURE 3–23   Select your color choices from the **Color Picker** box by clicking first on the Set Foreground Color in the floating toolbar.

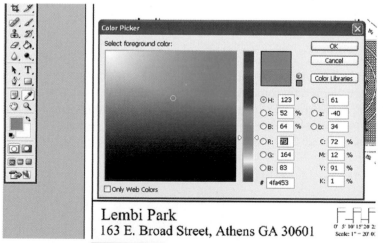

FIGURE 3–24   The **Color Picker** pop-up menu window options.

Once you select the color, make sure that the Color Render 1 layer is the active layer (highlighted in blue in the Layer Box window). Try to avoid a perfectly even color distribution, which will mimic air brush rendering technique more closely. Experiment with the different brush diameters, hardnesses, and opacities to get a softer or harder look. Imagine that the cursor arrow is the air brush nozzle, and use it to recreate the same movement and flow of the rendering media. You might consider overlapping a color over itself in order to get low and high values of color rendering (figure 3-25). There is no wrong and right in this technique since we are trying to recreate the feel of a hand drawing—not a perfect reproduction—in a mechanical rendering media.

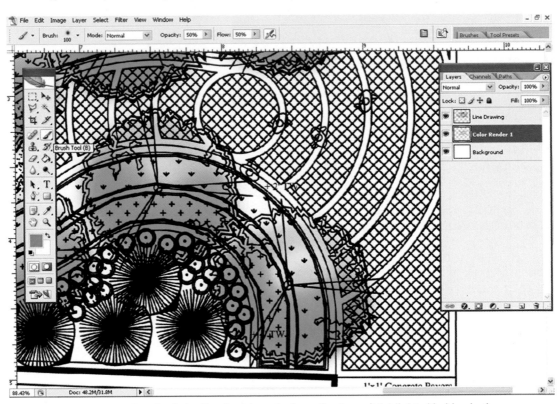

FIGURE 3–25   Make sure that the Color Render 1 layer is the active layer (highlighted in blue in the Layer Box window).

Do not hesitate to create more "Color Render" layers. This will allow you to overlap layers, thus preventing one layer from screening the other by accident (figures 3-26 and 3-27). Keep in mind that every time a new layer is created, by default Adobe Photoshop will make the new layer the top layer. Always move the Line Drawing layer to the top, thus preventing screening out the black lines with color. Use the Eraser tool to clean the overspills created by the air brush, or the Marquee tool to isolate specific areas to be rendered. Also, the Marquee tool allows pasting multiple copies of the same rendering such as trees or shrubs (figure 3-28).

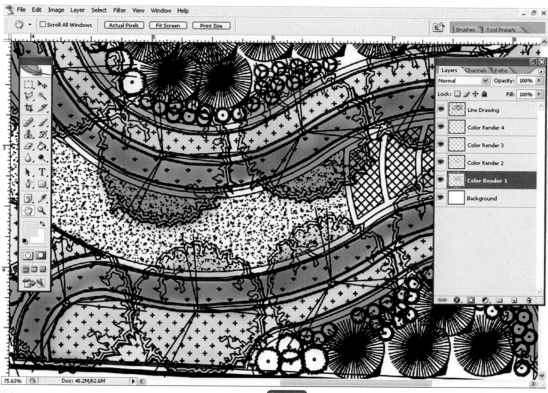

FIGURE 3–26 Create as many layers as needed in the **Layer** floating toolbar.

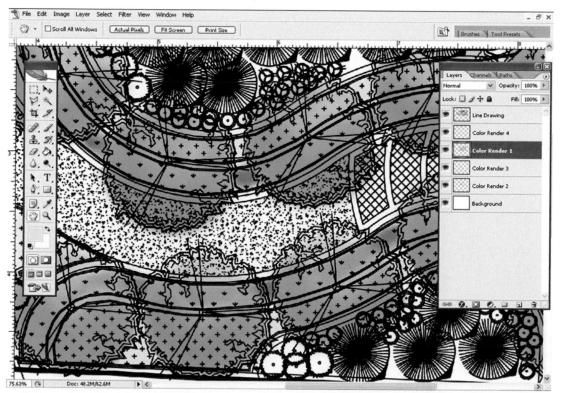

FIGURE 3–27 Change the layer order by selecting the layer (highlighted in blue), grab and pull to the top of the list.

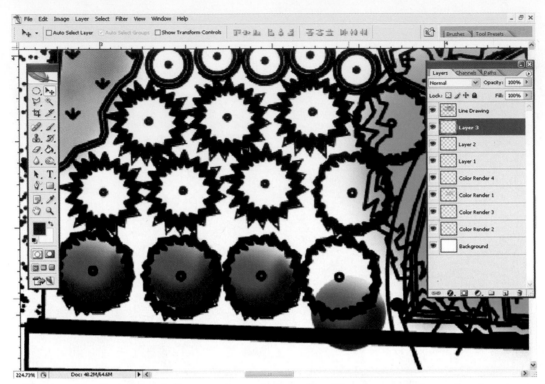

FIGURE 3–28 The Marquee tool allows pasting multiple copies of the same rendering such as trees or shrubs.

Adobe Photoshop by default action will create a new layer every time the `Copy` and `Paste` command is used. To save time, after completing the `Copy` and `Paste` task of the same object, turn all the layers off except the ones with the same object copied, and use the `Merge Visible` command under the `Layers` option of the top of the toolbar. This will merge all the active visible layers into one layer (figures 3-29 and 3-30).

Another time-saving option of Adobe Photoshop CS2 is that each layer opacity level can be manipulated to control the "transparency" of the layer. This will allow you to control whether different elements within the rendering composition are revealed or concealed from the view. Figures 3-31 and 3-32 show that, by changing the layer opacity of the Tree Rendering layer to a low setting, the understory flower beds can be revealed or concealed from the view.

Once you reach the desired "air brush" feel, save the drawing as an active Adobe Photoshop drawing and also save a copy as the "final drawing." The last one is the copy that you need to flatten all layers to create one simple drawing. The first one you keep as a backup copy, in case further revisions are needed (figures 3-32 and 3-33).

It is recommended that you print your drawing using Adobe PDF. Under `File` in the top toolbar, select `Save As` (figure 3-34). This will launch the `Save As` pop-up menu window. From there, scroll and select Photoshop PDF from the `Format` window. Make sure to name your drawing Color Print Format for easy recognition, and hit the `Save` button (figure 3-35). This will launch another `Save As` pop-up window. Hit the OK button until the last pop-up window is reached. Then, just click on the "Save PDF" button (figures 3-36, 3-37, and 3-38). This will turn the drawing into an easy format to print from `Adobe Acrobat Professional`, and will reduce the file size considerably without a noticeable loss of image resolution.

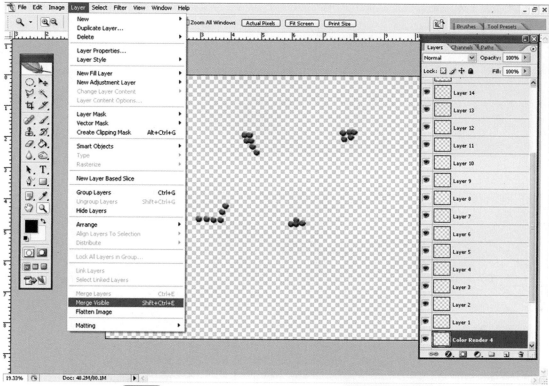

FIGURE 3–29 Under the **Layer** top toolbar, scroll down and select **Merge Visible**.

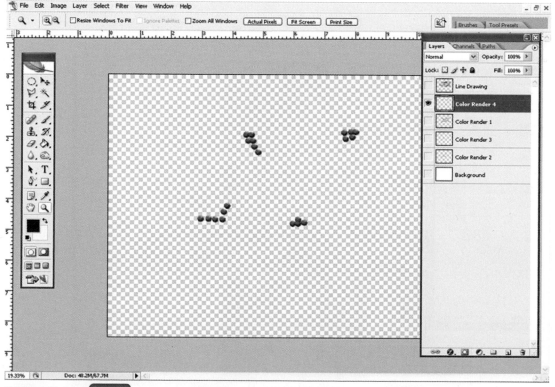

FIGURE 3–30 The **Layer** manager window shows all the visible layers now merged into one single layer.

FIGURE 3–31   Color Render 1 layer with opacity level set to 100%.

FIGURE 3–32   Color Render 1 layer with opacity level set to 26%.

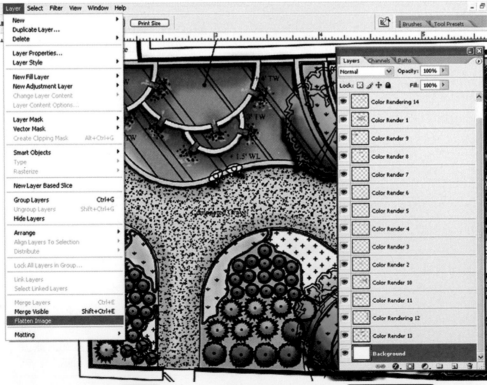

FIGURE 3–33   Under the Layer top tool tab, scroll down and select Flatten Image .

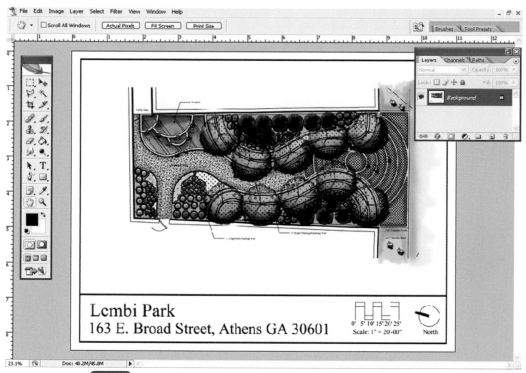

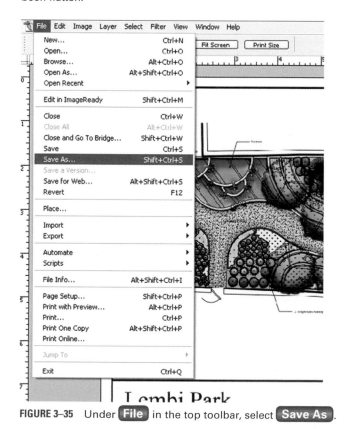

FIGURE 3–34 The Layer floating tool window showing only one layer (the active layer) after the image has been flatten.

FIGURE 3–35 Under File in the top toolbar, select Save As.

FIGURE 3–36 The **Save As** pop-up menu window scroll down menu option for image format.

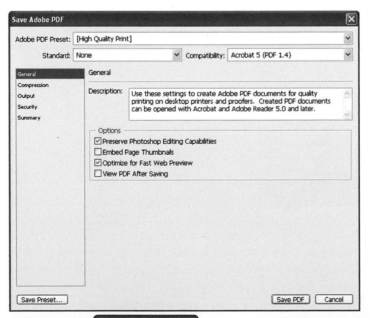

FIGURE 3–37 The **Save Adobe PDF** pop-up menu window preset to High Quality Print option.

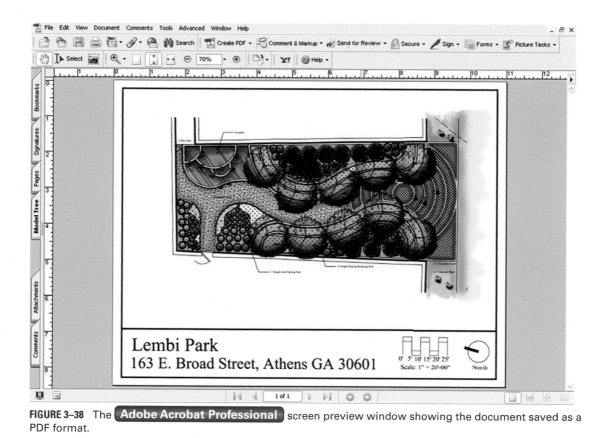

FIGURE 3–38   The **Adobe Acrobat Professional** screen preview window showing the document saved as a PDF format.

The "ACME" Blue Print Using Adobe Photoshop CS2

The blueprint format is imbedded in American pop culture thanks to two favorite cartoon characters, Wile E. Coyote and the Road Runner. The inventive Coyote was on a constant mission to catch the Road Runner, at one point devising and engineering a complex "ACME Blue Print Master Plan" to nab his elusive prey. This traditional image of the blueprint is especially attractive to those seeking a nostalgic feel in their drawings because it is a dated technology. Before the invention of the Xerox Laser Photocopying machine, reproductions of large-format plans were done with a Diazo Print Machine that used ammonia. This chemical reacted with light, turning the background of the paper dark blue, with high contrast white lines. Refinements in the Diazo Print Machine later reversed the color—thus the background of the paper turned a pale tone of blue (almost white) and the lines dark blue. Currently, because of environmental concerns and regulations, Diazo Print Machines are being replaced with ammonia-free oversized laser copy plotters and printers that can print in any color (including blue). Today it is the norm to get a good black-and-white copy, and the blueprint format is fading out of use. The nostalgic "old blueprint" format can still be created using Photoshop CS2, and is one of the simplest to accomplish if you follow these simple steps.

Setting the base drawing as a PDF format

The first step is to insert a good-quality black-and-white drawing into Photoshop CS2. There are several ways to insert a drawing into Photoshop CS2. The beginnings of Chapter 3 and Chapter 4 show how to insert a drawing into Adobe Photoshop as a PDF format. Also, the user can insert a scanned image (hand drawn) into Photoshop. Make sure to change the image format to a PDF before importing into Photoshop. The JPEG, TIFF, or **PSD** will not work with this technique. Also, make sure the imported PDF file is in black-and-white format in the RGB (red, green, blue) mode, and is of high resolution.

Setting the drawing

First, launch Adobe Photoshop CS2. Under **File** (top menu bar), select **Open**. This will launch another pop-up menu window in which the user can browse and locate the base drawing PDF format file. This will launch the **Import PDF** pop-up menu option window (figure 3-39). Make sure that the Crop To option is selected as Media Box, the Resolution is set to 150 or higher, and Mode is RGB. The original settings of the drawing, such as that of paper size, should be carried into Photoshop. If not, return to the original CAD base drawing and save again with the correct paper size proportions. This action will guarantee that the original scale of the drawings is preserved. Please refer to the previous section of this chapter for further information. By default setting, the PDF drawing will be placed in Layer 1. The checkerboard background indicates that the image is transparent except for the black lines of the original CAD PDF drawing. Make sure to select Fit On Screen under the top toolbar **View** icon for a better screen view work space setting, and dock the layer window to the side (figure 3-40).

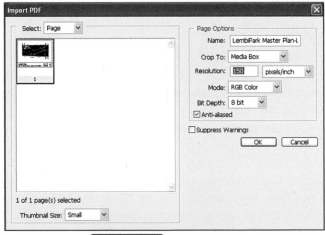

FIGURE 3–39   The **Import PDF** pop-up menu window.

At the **Layer** tool window, select Layer 1 and rename it Black Line Drawing. Also, create a new layer and name it Background. Make sure to place Black Line Drawing layer on top of the Background layer. This action will prevent accidentally screening the Black Line Drawing with the Background layer. Always label your layers for easy access and recognition (figure 3-41).

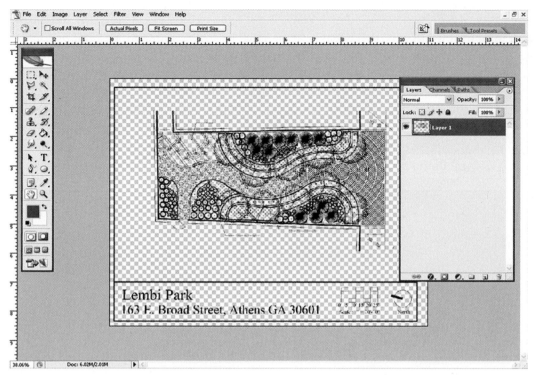

FIGURE 3–40 Adobe Photoshop CS2 screen view showing the imported PDF Lembi Park plan.

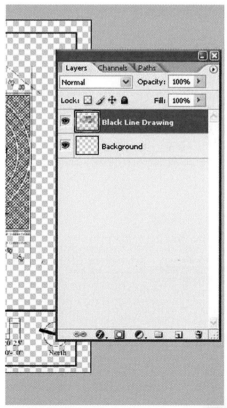

FIGURE 3–41 Rename the layers on the **Layer** floating tool window for easy recognition and visual reference per user needs.

Setting the dark blue background color

In the [Layer] tool window, turn off the Black Line Drawing layer and make the Background layer the active layer. Click on "Set Background Color," which is located in the side toolbar. This should launch the [Color Picker] window. Scroll up or down on the color bar to get the blue schemes, tones, and values. Select the blue color that is closest to the background blue color of an old blueprint, and then hit OK (figure 3-42).

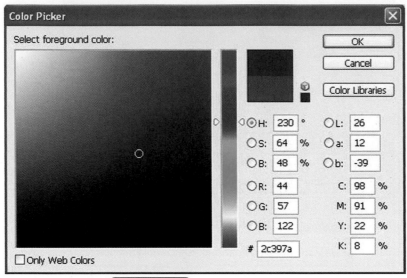

FIGURE 3–42   Use the [Color Picker] pop-up menu window to select the "blue print" background color.

From the side toolbar, select the Bucket tool and select the blue color from the background color, and click at the center of the checkerboard background. Make sure the Background layer is active. This action will fill the Background layer with a solid tone of blue color. Remember that the Background layer is meant to be under the Line Drawing layer, so make sure that your layer order is correct, otherwise the line drawing might end up underneath the solid field of blue. After the Background layer is turned dark blue, turn on the Black Line Drawing layer to double-check that your layer order is correct (figure 3-43).

Changing the black lines to white lines

First, make the Black Line Drawing layer the active layer. In order to change the lines from black/grey color, select the [Image] icon located at the top of the toolbar. Under [Image], select [Adjustments], and then select [Brightness/Contrast] (figure 3-44).

This will launch the [Brightness/Contrast] pop-up menu window. In this window change both field bars to +100. This will make the lines pure white (figure 3-45).

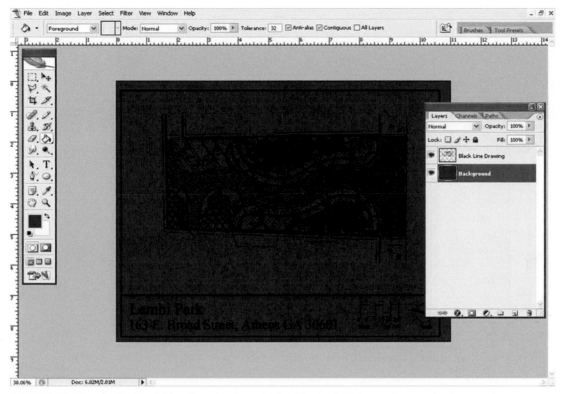

FIGURE 3–43 Turn on the Black Line Drawing layer to double-check that your layer order is correct.

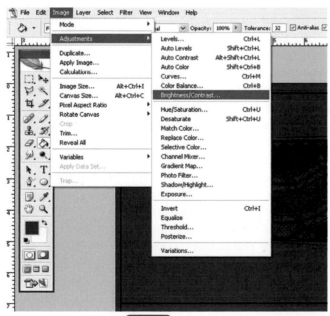

FIGURE 3–44 Under the **Image** top toolbar tab, scroll down and select **Adjustments**, and continue to scroll down on the menu, and finally select **Brightness/Contrast**.

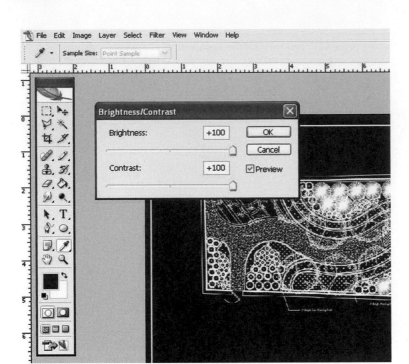

FIGURE 3–45 The Brightness/Contrast pop-up menu window.

Aging the blueprint

Anyone who ever came across an old blueprint plan will be familiar with fading blue edges, and water and stain marks on the paper. A combination of the Gradient, Dodge, and Burn tools can be used to achieve this old appearance. These tools will allow you to create a faded-out appearance on the background. Use the Gradient Editor and experiment with several color scheme combinations such as a dark and light tone of blue. The Gradient tool is located on the side toolbar. Look for the Bucket tool and click on the little corner arrow to launch the Gradient tool (figure 3-46). The gradient options are located at the top of the toolbar. Again, make sure the Background layer is the active layer. Use the Marquee tool to erase/delete the previous solid field of dark blue to start clean over a new background.

The Gradient tool option will allow the user to change the default settings of the highlights and dark values. Keep in mind that the default color for the gradient is the color that has been previously selected in the Color Picker, so make sure Dark Blue is selected before experimenting with the Gradient tool (figure 3-47).

To get a more aged appearance, use the Dodge or Burn tools to create highlights or dark spots along the edges or the center of the paper. Remember that all kinds of paper age with time, light, and humidity, and thus using the Dodge or Burn tool will accomplish this feel. Also, experiment with the style of brush, size, range, and exposure parameters. Use the "Drippy Water" brush from the Wet Media Brush styles in combination with the Dodge tool to create the effect of wet storage damage. Again, make sure the background (Dark Blue) layer is the active layer (figure 3-48).

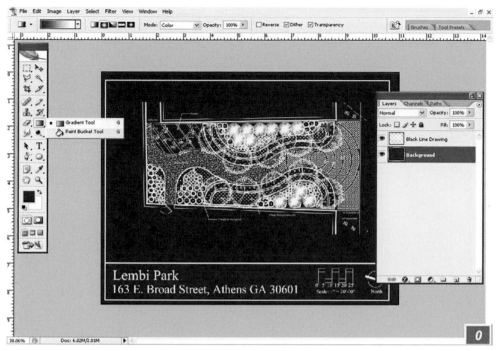

FIGURE 3–46 The Gradient tool option is located in the floating toolbar, next to the Eraser Icon tool.

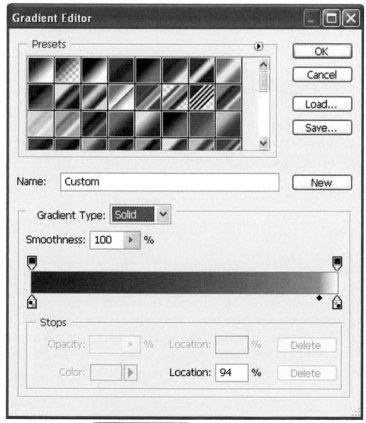

FIGURE 3–47 The **Gradient Editor** pop-up menu window.

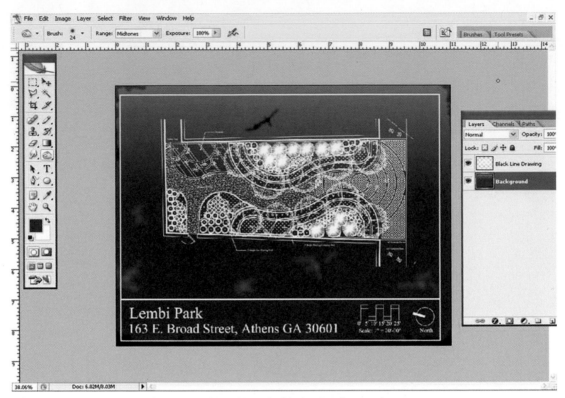

FIGURE 3–48 Faded blue background color achieved with the Gradient tool.

Saving the image and plotting

After completing the old blueprint look, save two copies of the drawing. Save one copy as an active Adobe Photoshop drawing and the second as the finished image. The finished image copy is the one that you need to flatten all layers to create one simple drawing. The active Adobe Photoshop drawing copy you will keep as a backup copy in the event that further revisions are required. **Print/Save** your flattened finished copy drawing as an Adobe PDF format. The PDF format is printer and electronic transfer friendly, and also reduces the file size (storage capacity) considerably.

The Standard Sepia Color Print

Like the blueprint, the sepia print is another print format based on the Diazo Print Machine technology of the past. The sepia print's main characteristic is that the lines are dark brown and the background color is a pale coffee tone. The sepia print was also favored because the dark brown lines are erasable. Also, the paper is almost translucent, thus allowing for revisions directly on the sepia copy and then running the sepia print again through the Diazo Print Machine to make more copies. Like the blueprint, the sepia offers an old-fashioned and nostalgic look that is easily reproduced using Adobe Photoshop CS2.

Setting the drawing

Using exactly the same steps previously described for setting the blue print, import a CAD base drawing into Adobe Photoshop CS2 as a PDF file. Likewise, create a new layer and label it Background. Rename the PDF drawing layer as Black Line Drawing. As previously stated, this simple step allows us to quickly recognize where things are located. Next, make the Background layer the active layer and turn off the Line Drawing layer (figure 3-49).

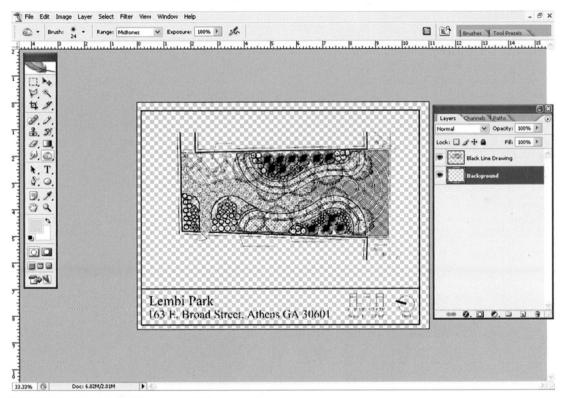

FIGURE 3–49 The Adobe Photoshop CS2 standard screen.

Setting the light tan background color

Following the same steps as previously explained in the blue print format, in the **Layer** tool window, turn off the Black Line Drawing layer, and make the Background layer the active layer. Using the **Color Picker** tool, select a light tan color that is similar to an old sepia color, then using the Bucket tool fill the background color with your sepia color choice (figures 3-50 and 3-51).

This action will fill the Background layer with a solid tone of tan color. Remember that the Background Layer is under the Line Drawing layer, so make sure that your layer order is correct, otherwise the line drawing might end up underneath the solid field of tan.

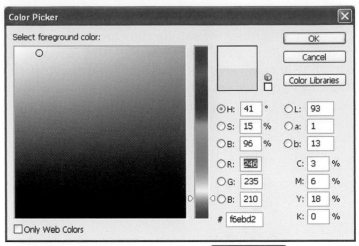

FIGURE 3–50   Setting the tan color on the Color Picker pop-up menu window.

FIGURE 3–51   Create a faded/antique-feel background color with the Gradient tool.

Changing the black lines to dark brown

To change the black lines to dark brown, you must first make the Black Line Layer the active layer. Next, at the top of the toolbar select Image . Under Image , select Adjustments , followed by Hue/Saturation (figure 3-52). This action will launch the Hue/Saturation pop-up menu window (figure 3-53).

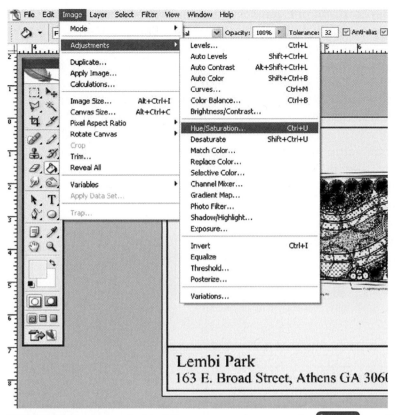

FIGURE 3–52 To change the black lines to tan or brown, under **Image**, scroll down and select **Adjustments**, and continue to scroll down and finally select **Hue/Saturation**.

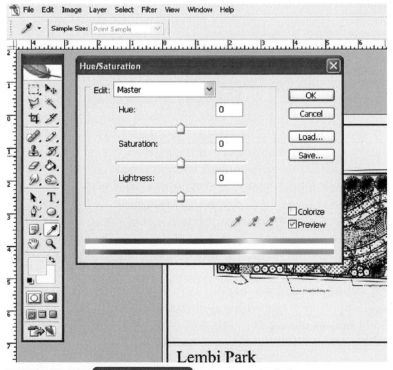

FIGURE 3–53 The **Hue/Saturation** pop-up menu window.

Next, use the Eyedropper tool and click/select the black line of the drawing. Once you select the black line of the drawing, notice that in the Select Background/Foreground color tool, located in the side toolbar, the square preview turns black. Then, on the **Hue/Saturation** pop-up menu window, check the Colorize and **Preview** icons boxes. In this pop-up window change the Hue Field box option to 34, the Saturation box option to 100, and the Lightness box option to 24. This will change the color of the lines from black to brown (figure 3-54). Note that the Select Background/Foreground window of the side bar will change from black to selected tone of brown. Also, keep in mind that depending on your monitor color setup, the box options of the Hue, Saturation, and Lightness may differ from the ones stated above. You may need to try different values on the box options to get the desired tones of brown. Once you reach your brown color, hit the OK button to set (figure 3-55).

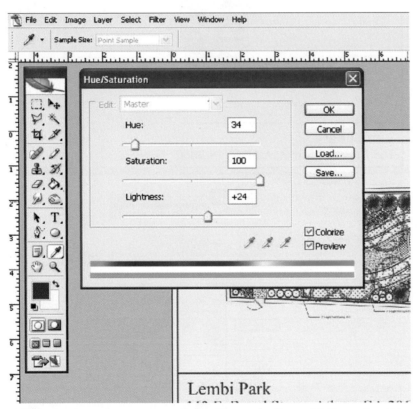

FIGURE 3–54 To achieve the tan or brown line set the Hue to 34, Saturation 100, and Lightness to 24.

Aging the sepia print

Aging the sepia print plan will reinforce the nostalgic look and feel of the plan to resemble this dated technology. Like in the previous blue print exercise, use the Gradient tool to create the illusion of paper edge fading (figure 3-56). The Gradient tool is located in the side toolbar. Click on the little corner arrow in the Bucket tool icon in order to select the Gradient tool. To launch the **Gradient Editor** pop-up menu window, double-click on the Gradient sample bar located on the top toolbar of your screen.

FIGURE 3–55   Adobe Photoshop CS2 screen showing the finish sepia print look.

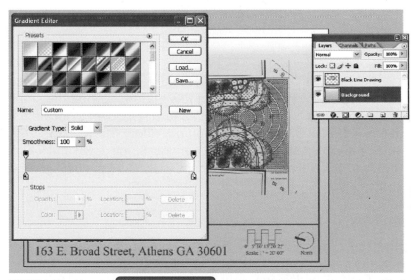

FIGURE 3–56   Use the **Gradient Editor** pop-up menu window to create a faded look.

Another way to reinforce the aged look is by using the Dodge or Burn tools. To further edit the Dodge or Burn tool options, double-click on the small arrow next to the brush size of the top toolbar of your screen. This will launch the **Brush Editor** pop-up menu window (figure 3-57).

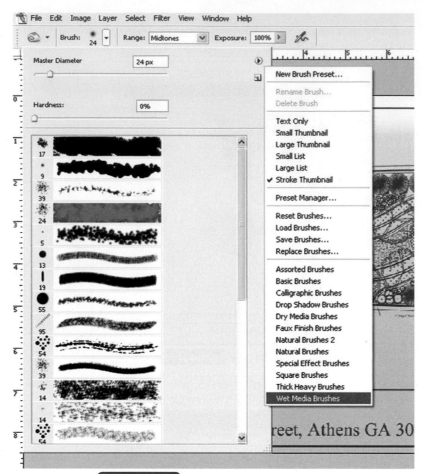

FIGURE 3–57   The [Brush Editor] pop-up menu window.

Try different brush sizes, shapes, and styles. The [Wet Media Brushes] will reinforce the illusion of wet storage damage on the paper. To select this option, click on the small arrow on the top of the [Brush Editor] pop-up menu window and scroll down to locate the [Wet Media Brush] option (figure 3-57). Hit OK in the next pop-up window. The [Brush Editor] pop-up menu window will now show the new/loaded brush options. Scroll down and locate the Drippy Water brush and click to select it (figure 3-58). Again, make sure the Background (light tan color) layer is the active layer.

Saving the image and plotting

As in the previous blue print exercise, after completing the sepia print look, save two copies of the drawing. Save one copy as an active Adobe Photoshop drawing and the second as the finished image. The finished image copy is the one that you need to flatten all layers to create one simple drawing. The active Adobe Photoshop drawing copy you will keep as a backup copy in the event that further revisions are required. [Print/Save] your flattened finished image copy drawing as an Adobe PDF format. The PDF format is printer and electronic transfer friendly, and also reduces the file size (storage capacity) considerably (figure 3-59).

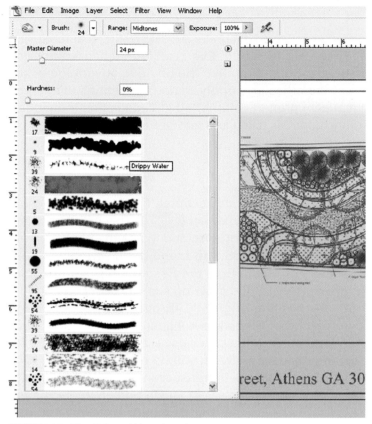

FIGURE 3–58 The Drippy Water brush.

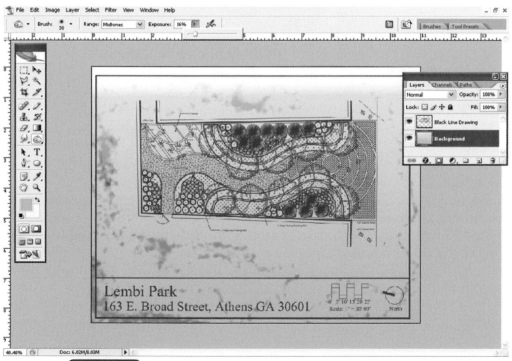

FIGURE 3–59 Print/Save your flattened finished image copy drawing as an Adobe PDF format.

Water Color

Today's landscape architecture computer-savvy practitioners are discovering that rendering in any kind of media is possible in the digital world. Even traditional hand-rendering media such as water colors can be recreated digitally. The obvious advantage of the savings in annual overhead cost of the office art supply materials justifies the investment in a laptop computer and a quality printer. Another advantage is that individuals who are allergic to certain art supply materials, such as acetone-based markers, are now able to express their artistic capabilities digitally.

Setting the base drawing as a PDF format

As previously explained in the blue print section of this chapter, first start by launching Adobe Photoshop CS2. Under **File** (top menu bar), select **Open**. This will launch another pop-up menu window in which the user can browse and locate the base drawing PDF format file. This will launch the **Import PDF** pop-up menu option window. Make sure that the Crop To option is selected as Media Box, the Resolution is set to 150 or higher, and Mode is RGB. The original setting of the drawing such as paper size should be carried into Photoshop. If not, return to the original CAD base drawing and save again with the correct paper size proportions. This action will guarantee that the original scale of the drawing is preserved. Please refer to the previous section of this chapter for further information. By default setting, the PDF drawing will be placed in Layer 1. The checkerboard background indicates that the image is transparent except for the black lines of the original CAD PDF drawing. Make sure to select Fit On Screen under the top toolbar **View** icon, and dock the layer window to the side for a better screen view work space setting.

At the **Layer** tool window, select Layer 1 and rename it Black Line Drawing. Also, create a new layer and name it Background. Make sure to place Black Line Drawing layer on top of the Background layer. This action will prevent accidentally screening the Black Line Drawing with the Background layer. Always label your layers for easy access and recognition (figure 3-60).

Setting the background color

In the **Layer** tool window, turn off the Black Line Drawing Layer and make the Background the active layer. Click on "Set Background Color," which is located in the side toolbar. This will launch the **Color Picker** window. Scroll up or down on the color bar to get the background color of your choice, and then hit OK.

From the side toolbar, select the Bucket tool and select your color from the background color; next click at the center of the checkerboard background (figure 3-61). Make sure the Background layer is active. This action will fill the Background layer with a solid tone of the color of your choice. Remember that the Background layer is under the Line Drawing layer, so make sure that your layer order is correct—otherwise the line drawing might end up underneath the solid field of your chosen color. After the Background layer is set, turn on the Black Line Drawing layer to double-check that your layer order is correct.

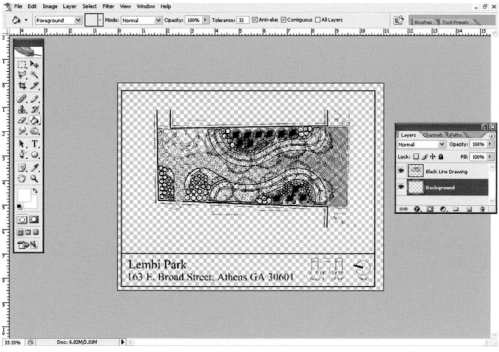

FIGURE 3–60 The Adobe Photoshop CS2 standard screen.

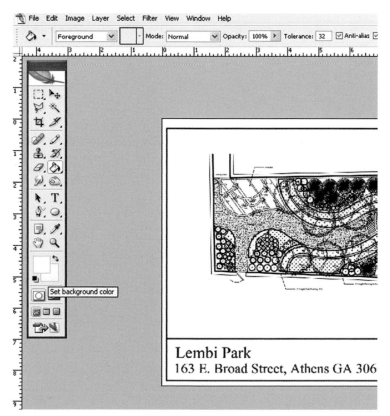

FIGURE 3–61 Access the Set Background Color tool by clicking on the icon on the floating tool window.

author's note

Although solid off-white or light pastel colors are the most common choices for water color papers, they come in a full spectrum of colors. Choose the background color that best suits your color scheme needs.

Setting the layers and water color style brushes

Adobe Photoshop CS2 and AutoCAD offer the advantage of working in layers. The layers offer the capability of isolating items within the composition and the flexibility of overlapping/screening. The next step is to create several layers between the Line Drawing and Background layer. The Black Line Layer must always be in the top of the **Layer Editor** window to prevent the black lines from being covered by color. The Background layer must always be in the bottom of the **Layer Editor** window to prevent the solid field of color from accidentally screening everything before flattening the image. Start by creating several new layers, placing them between Black Line and Background layers. Label these new layers Color Render 1, 2, 3, and so forth (figure 3-62).

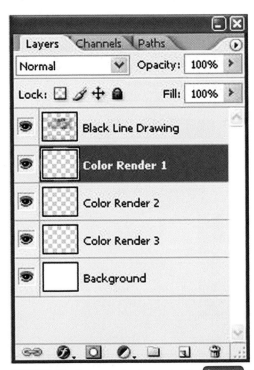

FIGURE 3–62 Rename the layers in the **Layer** floating window.

Make the Color Render 1 layer the active layer. Next, select the Brush tool located in the side toolbar. Click on the corner arrow of the Pencil icon to select the Brush tool (figure 3-63).

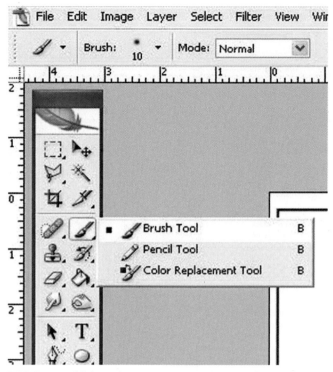

FIGURE 3–63 Click on the corner arrow of the Pencil icon to select the Brush tool.

Once the Brush tool is selected, the top of the toolbar will show the Brush Editing options, such as Brush Size, Mode, Opacity, and Flow. Click on the arrow next to the Brush Size to launch the Brush Editor pop-up menu window. Next to the Brush Diameter, click on the arrow to open the pop-up menu list for Brush Media option, scroll down, and select Wet Media Brushes (figure 3-64).

The Brush Editor Preview window will now show the loaded Wet Media Brush options. Scroll down and select from the list the Drippy Brush, and change the Brush Master Diameter and Hardness to the parameter that best suits your needs (author's choice is 10 and 50). Once selected, look at the top of the toolbar of your screen and change the Mode to Normal, the Opacity to 50%, and the Flow to 50%. This will allow you to create the transparent and soft look of real water color (figure 3-65).

Color rendering

Next, select your color choices from the Color Picker box, and hit OK (figure 3-66). Make sure that Color Render 1 layer is the active layer. Also, click on the Brush tool icon and then use the mouse to apply color into the drawing. Think of the mouse as a brush. Do not try to get a perfectly even color distribution, as water color renderings have the tendency to be uneven. Also, do not hesitate to change the brush diameter, hardness, and opacity to get a softer or harder look. Using the mouse as a brush, try to recreate the same movement and direction of an actual brush. You might consider overlapping a color over itself to get low and high values of color rendering. There is no wrong and right in this technique since we are trying to recreate the feel of an imperfect, hand-rendered media. Also, use the Zoom tool to fit the object in the screen and enable to see the entire rendering (figure 3-67).

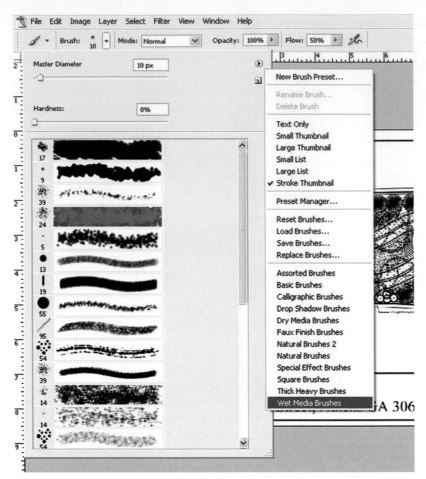

FIGURE 3–64 Click on the arrow to open the pop-up Menu list for Brush Media Option, scroll down and select ▕ **Wet Media Brushes** ▏.

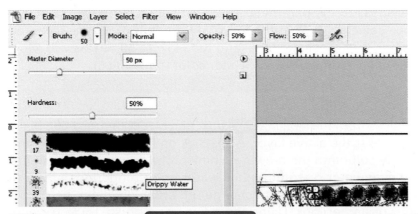

FIGURE 3–65 From the new ▕ **Wet Media Brush** ▏ styles, scroll down and select the Drippy Water brush.

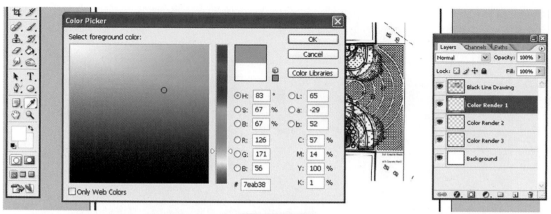

FIGURE 3–66   Select your color choice from the ⬤ Color Picker ⬤ pop-up menu window.

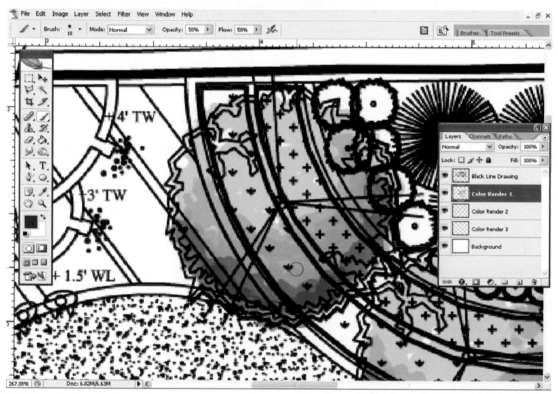

FIGURE 3–67   Use the Zoom tool to fit the object in the screen and enable to see the entire rendering.

Do not hesitate to create more "Color Render" layers. This will allow you to overlap one layer over the other, thus preventing one layer from screening the other by accident. Use the Eraser to clean some of the overspills created by the Drippy Brush, or use the Marquee tool to isolate specific areas to be rendered (figure 3-68).

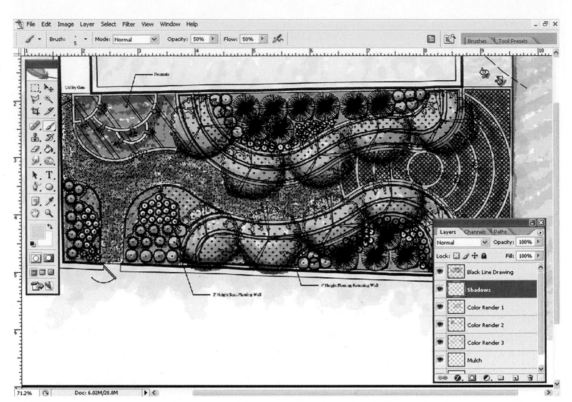

FIGURE 3–68 Use the eraser to clean some of the overspills created by the drippy brush, or use the Marquee tool to isolate specific areas to be rendered.

Once you render your composition, you need to "texturize" all the color layers in order to enhance the water color paper feel.

Merging the layers and texturizing the water color paper

Most water color paper has some kind of texture. Adobe Photoshop CS2 will allow you to texture your drawing to match that of wet media paper, thus enhancing the water color feel of your rendering. In order to "texture" your drawing, the **Image** mode must be set in **RGB Color** to 8 Bits per channel (figure 3-69).

The Texturizing tool will complete the feel and appearance of water color. This tool will create a visual texture that resembles canvas, burlap, sandstone, and brick. Before using this tool, save one document copy first as an active Photoshop document, and a second copy to be used to create the finished image. The active Photoshop document copy you will keep as a backup copy in case further revisions are required. The second copy is the one to merge all the color layers and the background first before texturize. Keeping an active photoshop document with all the layers intact will enable the user to make further changes or revisions to the original. Once the layers are flattened, no further manipulation or changes are allowed. Do not texturize the Black Line layer because this might create some distortion, thus weakening the general appeal of the composition. To avoid Texturizing the Black Line layer, turn this layer off first before merging all the layers and texturizing (figure 3-70).

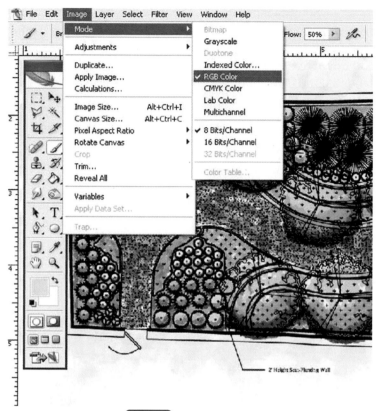

FIGURE 3–69 Under the **Image** top toolbar, scroll down to Mode and select **RGB Color**.

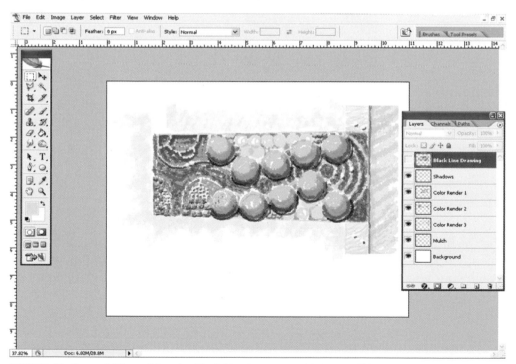

FIGURE 3–70 To avoid Texturizing the Black Line layer, turn this layer off first before merging all the layers and texturizing.

Once you turn the Black Line layer off, make the Background layer active. Also, make sure the layer order is set correctly since the layers will be flattened one on top of the other per the order that they are placed in the **Layer** window. At the top of the toolbar and under **Layer**, click and scroll down to select **Merge Visible** (figure 3-71). All the visible layers are now bound together as one layer (flattening). When an image is flattened, the name of the layer that is active will be preserved. **Merge Visible** tool will flatten only the visible (On) layers. Because the Dark Line layer was turned off first, this layer will remain intact (figure 3-72).

The next step is to texturize the Background layer only. At the top toolbar, under **Filter**, scroll down and select **Texture**, and then select **Texturizer** (figure 3-73). The **Texturizer Editor** pop-up window will be launched (figure 3-74). The **Texturizer Editor** box will have several options to choose from. Select the one that best suits your aesthetic needs. Also, explore modifying the scale, relief, and light source. Hit OK when the desired texture is attained.

FIGURE 3–71 **Merge Visible** tool will flatten only the visible (On) layers.

Once the OK button is hit, the **Texturizer Editor** window will close, returning to the original screen. Turn the Black Line layer on to see the finished composition (figure 3-75). Once you reach your finish rendering look, flatten all the remaining layers before plotting.

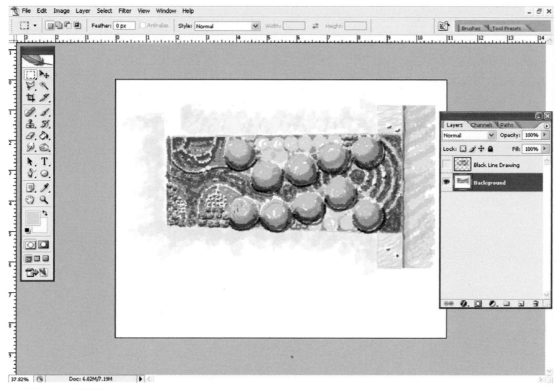

FIGURE 3–72 All "ON" color layers are merged together in one.

FIGURE 3–73 Under the **Filter** tool tab, scroll down to **Texture** and select **Texturizer**.

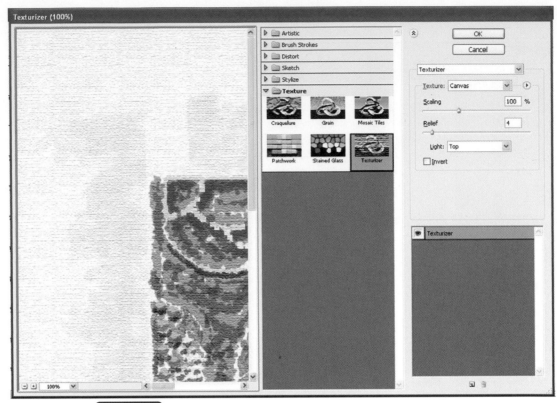

FIGURE 3–74 The Texturizer pop-up menu window options.

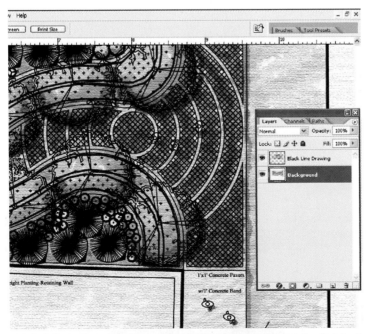

FIGURE 3–75 Zoom into the image to see and appreciated the finished product.

Saving the image and plotting

As in the previous blue print and sepia print exercises, after completing your water color rendering, save two copies of the drawing. Save one copy as an active Adobe Photoshop drawing and the second as the finished image. The finished image copy is the one that you need to flatten all layers to create one simple drawing. The active Adobe Photoshop drawing copy you will keep as a backup copy in the event that further revisions are required. **Print/Save** your flattened finished image copy drawing as an Adobe PDF format. The PDF format is printer and electronic transfer friendly, and also reduces the file size (storage capacity) considerably (figure 3-76).

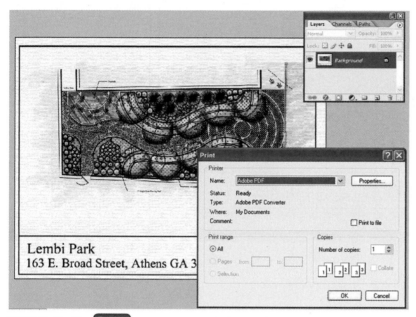

FIGURE 3–76   The **Print** pop-up menu window option.

Marker Rendering

Adobe Photoshop CS2 will enable users to recreate the colors, feel, and texture of markers. Rendering with color markers has become the signature hand rendering technique in today's professional practice of landscape architecture. Color markers were first introduced in the 20th century as a technological advancement and substitute for the traditional media of water color paint. The color markers offered the advantages of being easy to transport, fast drying, and inexpensive. They also came in a variety of colors, didn't require water color paper, and were less messy than water color paint. Even though color markers are a dated color rendering technique, they still are one of the most important and flexible illustration tools in the profession.

Setting the base drawing as a PDF format

To import a CAD base drawing as a PDF file into Adobe Photoshop CS2, follow the same steps as previously explained for setting the water color base drawing. Again, make sure to properly import your CAD drawing as a PDF file into Adobe Photoshop. Also, rename and create several layers (Black Line Drawing layer, Color Render layer, and Background) per the user needs. Make the Background layer a solid field of white for better color contrast and readability (figure 3-77).

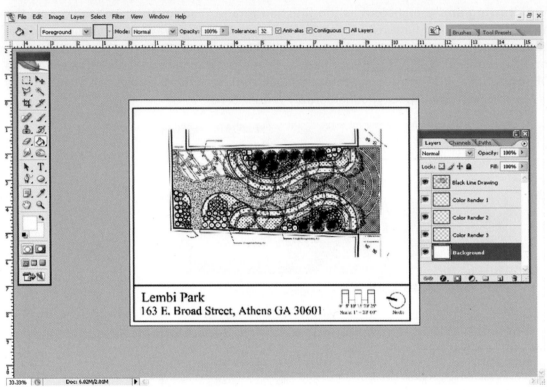

FIGURE 3–77 The Adobe Photoshop CS2 standard window screen.

Setting the "Marker" Brush Style

The Brush tool will enable the user to create the feel and texture of the marker in Adobe Photoshop CS2 by modifying the **Brush Editor**. First, make sure the Color Render layer is active. Then, select the Brush tool from the side toolbar of the screen. Please note that the top of the toolbar of your screen now shows the Brush tool editing options of Brush Tool Type, Brush Tool Preset Picker, Mode, Opacity, and Flow (figure 3-78).

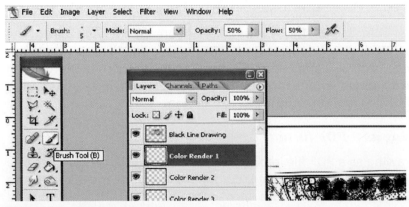

FIGURE 3–78 Select the Brush Tool icon on the floating toolbar.

On the top toolbar of the Brush tool option, click on the arrow next to the first Brush icon. This will open the (Brush Type) pop-up Menu options; now scroll down the list and select the Medium Marker Tip (figure 3-79). This option is the closest to the traditional broad-chisel soft tip of a color marker. Next, click to open the Brush Tool Preset Picker, and change the size of the marker tip to match the actual size of a real marker chisel tip (about ± ¼″). Set the Master Diameter from 30 to 55 pixels maximum (figure 3-80). Use the ruler on the top or side of the document page as a guide to get an approximate size. Also, keep the Mode in Normal, and change the Opacity to 50% and the Flow to 50%. The next step is to select your color choice from the (Color Picker) box.

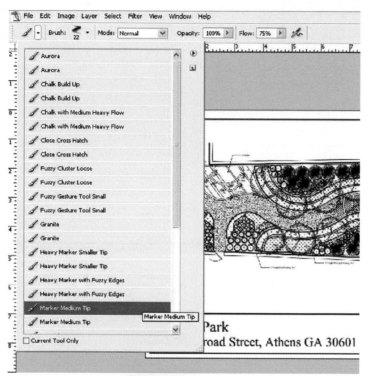

FIGURE 3–79 Under the (Brush Type) Popup Menu options, scroll down the list and select the Medium Marker Tip.

Color rendering

Once you set the (Brush Editing) options to match the feel and texture of a marker, select a color from the (Color Picker) box (figure 3-81). Make sure the Color Render layer is the active layer before rendering with the Brush tool. Experimentation is the key in this technique, so use the mouse like a marker and try different angles and strokes. Remember that markers do not create a perfectly even color distribution; try leaving white gaps between angles and strokes, and overlap the same color to get a darker tone and value. There is no wrong and right in this technique, as we are trying to recreate the "sloppy" marker media. Create more "Color Render" layers as needed (do not put everything in one layer). This will allow you to overlap one layer over the other, thus preventing one layer from screening the other by accident. Uses the Eraser to clean some of the over-spills created by the marker strokes, or use the Marquee tool to isolate specific areas to be rendered. Also, use the Zoom tool to fit the object in the screen and enable to see the entire rendering of small areas (figure 3-82).

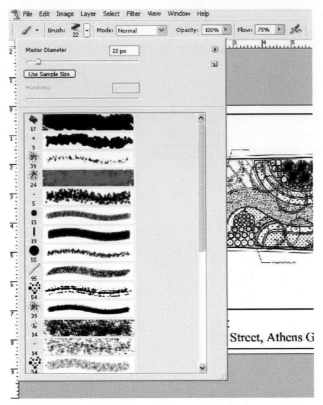

FIGURE 3–80 Set the Master Diameter from 30 to 55 pixels maximum.

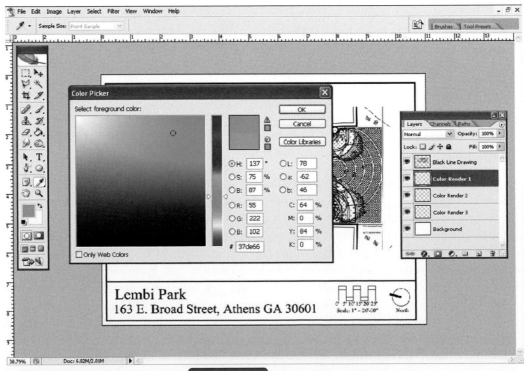

FIGURE 3–81 Select a color from the **Color Picker** pop-up menu window.

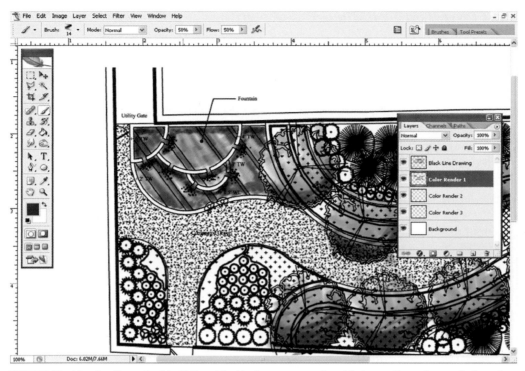

FIGURE 3–82   Use the Zoom tool to fit the object in the screen and enable to see the entire rendering of small areas.

Once you render the entire composition, add highlights and shadows to enhance the depth of the drawing at the end. Create a shadow layer and place it below the Black Line layer in order for the shadows to cover all colors under it (figure 3-83).

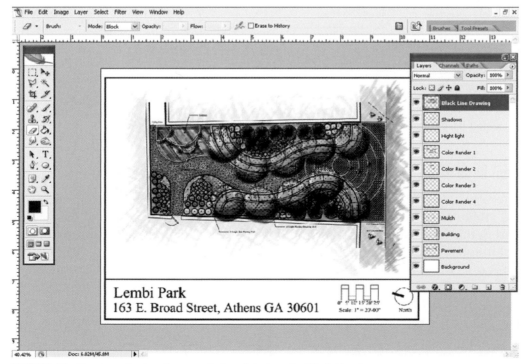

FIGURE 3–83   Create a shadow layer and place it below the Black Line layer.

Merging the layers, saving the image, and plotting

As in the previous blue print, sepia print, and water color exercise, after completing your marker color rendering, save two copies of the drawing. Save one copy as an active Adobe Photoshop drawing and the second as the finished image. The finished image copy is the one that you need to flatten all layers to create one simple drawing (figure 3-84). The active Adobe Photoshop drawing copy you will keep as a backup copy in the event that further revisions are required. **Print/Save** your flattened finished image copy drawing as an Adobe PDF format (figure 3-85). The PDF format is printer and electronic transfer friendly, and also reduces the file size (storage capacity) considerably (figure 3-86).

FIGURE 3–84   Under the **Layer** top toolbar tab, scroll down and select **Flatten Image**.

Color Pencil Rendering

Color pencil rendering is another standard technique in today's professional practice of landscape architecture. Like color marker rendering, this rendering technique was first introduced in the 20th century, as a technological advancement and substitute for the traditional media of water color paint and ink. It offered the same advantages of color markers but on a dry format. Adobe Photoshop CS2 will enable the user to recreate the feel and texture of this traditional hand rendering media by virtue of a combination of "Brush and Color Picker" tools.

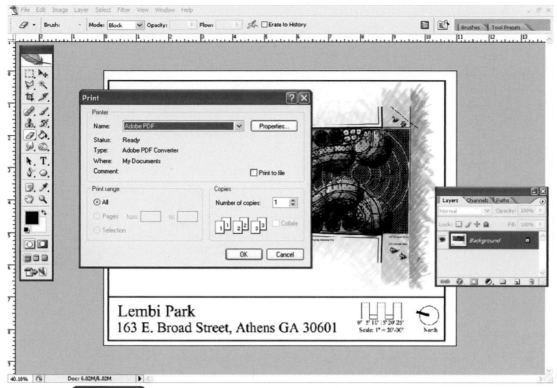

FIGURE 3–85 **Print/Save** your flattened finished image copy drawing as an Adobe PDF format.

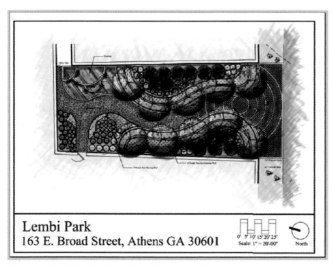

FIGURE 3–86 The finished PDF file drawing.

Setting the CAD base drawing as a PDF format

Readers should be familiar with the steps required to import a CAD base drawing as a PDF file into Adobe Photoshop CS2—follow the same steps as previously explained for setting the water color base drawing. Again, make sure to properly import your CAD drawing as a PDF file into Adobe Photoshop. Also, rename and create several layers

(Black Line Drawing layer, Color Render layer, and Background) per the user needs. Make the Background layer a solid field of white for better color contrast and readability (figure 3-87).

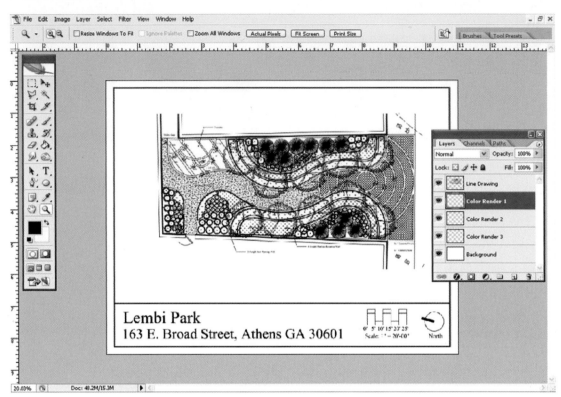

FIGURE 3–87   The Standard Adobe Photoshop CS2 window screen.

Setting the "Pencil" Brush Style

The Brush tool will enable the user to create the feel and texture of the color pencil in Adobe Photoshop CS2 by modifying the **Brush Editor**. First, make sure the Color Render layer is active. Then, select the Brush tool from the side toolbar of the screen. Please note that the top of the toolbar of your screen now shows the Brush tool editing options of Brush Tool Type, Brush Tool Preset Picker, Mode, Opacity, and Flow (figure 3-88).

On the top toolbar of the Brush tool option, click on the arrow next to the first Brush icon. This will open the **Brush Type** pop-up menu options. Click on the arrow on the top of this pop-up menu window, scroll down the list, and select the **Dry Media Brushes** (figure 3-89). This will load the "preset" Brush Style options for **Dry Media Brushes** in the **Preview** window. Scroll down this list and select the #2 Pencil option (figure 3-90). This option comes closest to the traditional feel and texture of color pencils. At this point, keep the Mode in Normal, the Opacity to 50%, and the Flow to 50%. The next step is to select your color choice from the **Color Picker** box.

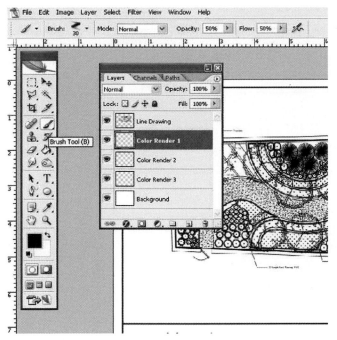

FIGURE 3–88   The top toolbar of your screen now shows the Brush Tool Editing options.

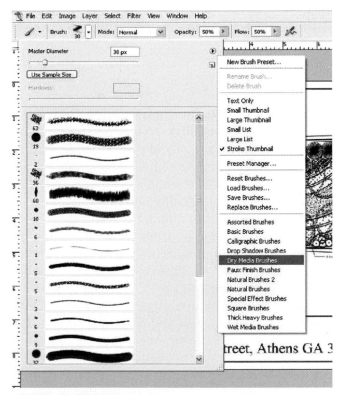

FIGURE 3–89   Click on the arrow on the top of this popup menu window, scroll down the list, and select the **Dry Media Brushes**.

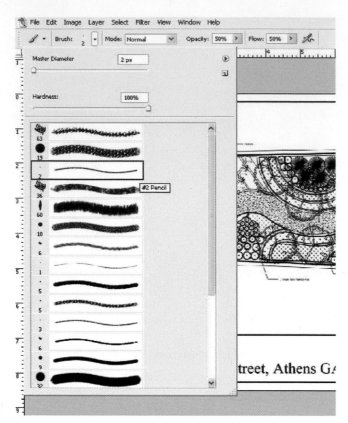

FIGURE 3–90 Scroll down this list and select the #2 Pencil option.

Color pencil rendering

Next, select your color choices from the Color Picker box, and hit OK (figure 3-91). Make sure that Color Render 1 layer is the active layer. Also, click on the Brush Tool icon and then use the mouse to apply color into the drawing. Think of the mouse as a color pencil. Do not try to get a perfectly even color distribution since pencil color renderings have the tendency to be uneven. Also, do not hesitate to change the "Brush Tool" diameter, hardness, and opacity to get a softer or harder look. Imagine that the mouse is a color pencil and try to recreate the same movement and direction of an actual color pencil. You might consider overlapping a color over itself in order to get low and high values of color rendering. There is no wrong and right in this technique as we are trying to recreate the feel of a hand-rendered media. Also, use the Zoom tool to fit the object in the screen and enable to see the entire rendering (figure 3-92).

Color pencil rendering's signature is the "rogue" texture, angle of the strokes, and overlapping of different tones and values that can be seen and read at a simple distance (figure 3-93). You can try keeping all the pencil strokes in one direction, or even a crosshatch pattern. Create more "Color Render" layers as needed. This will allow you to overlap one layer over the other, thus preventing one layer from screening the other by accident. Use the Eraser to clean some of the overspills created by the Pencil, or use the Marquee tool to isolate specific areas to be rendered or to copy and paste multiple patterns (figure 3-94).

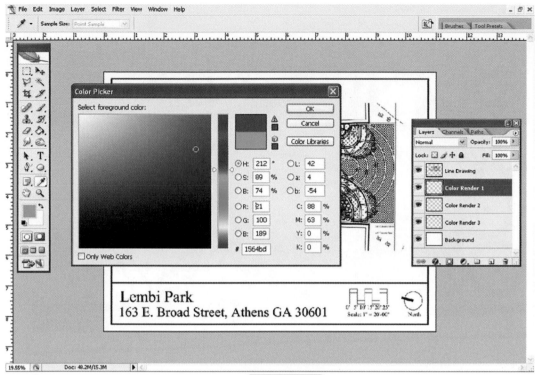

FIGURE 3–91 Select your color choices from the **Color Picker** box.

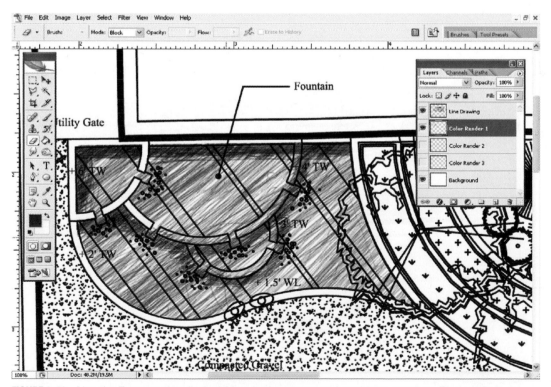

FIGURE 3–92 Use the Zoom tool to fit the object in the screen and enable to see the entire rendering.

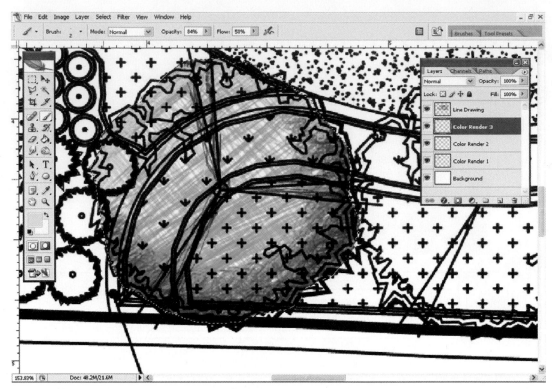

FIGURE 3–93 Closeup view of the color pencil rendering signature texture.

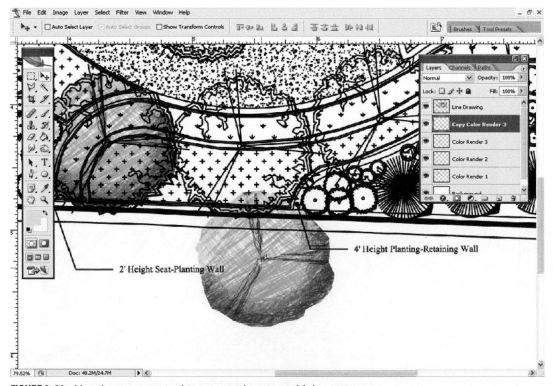

FIGURE 3–94 Use the marquee tool to copy and paste multiple patterns.

Rendering shortcuts

Color pencil rendering might be a little time consuming for large rendering areas such as asphalt, sidewalks, lawns, lakes, and buildings. One way to speed up the rendering process is to create a basic pattern that can be copied and pasted as many times as needed. The trick is to create a pattern that can be seamlessly pasted together, thus creating the illusion of one large unit. In the case of this chapter illustration for color pencil rendering, the color pencil strokes are done in 45° angles. One way to achieve this basic pattern is to first make one of the Color Rendering layers the active layer and then to use the Polygonal Lasso Tool to create a rectangle area at 45° angles (parallelogram shape) (figure 3-95).

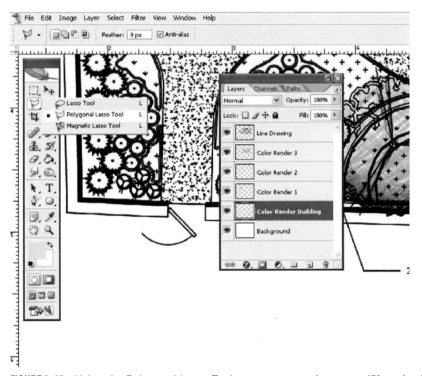

FIGURE 3–95 Using the Polygonal Lasso Tool create a rectangle area at 45° angles (parallelogram shape).

Draw inside your parallelogram shape (Polygonal Lasso) area the pencil lines following a 45° angle (figure 3-96).

Using the **Edit** – **Copy** and **Edit** – **Paste** tools, create multiple copies of the Parallelogram pattern area (figures 3-97 and 3-98).

Use the Move tool to move and stagger each copy of the pattern area in a seamless fashion (figure 3-99).

Users are aware that the default settings of the copy and paste tools will create a new layer for each copy object in the **Layer Management** window. Once your area is complete with the pattern, merge all the copied layers into one. Before using the **Merge Visible** tool make sure to first turn off all the other layers that are not the parallelogram pattern (figure 3-100).

Once all the pattern layers are merged together, use the Polygonal Lasso tool or Eraser to delete and clean up all the overspills (figure 3-101).

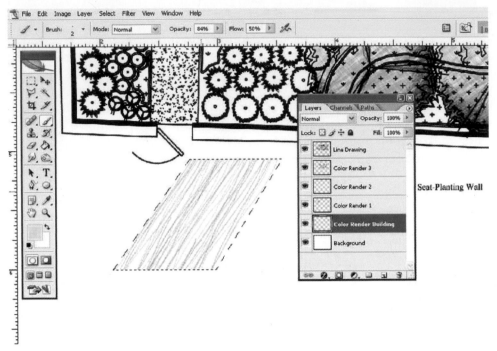

FIGURE 3–96 The pencil lines follow a 45° angle.

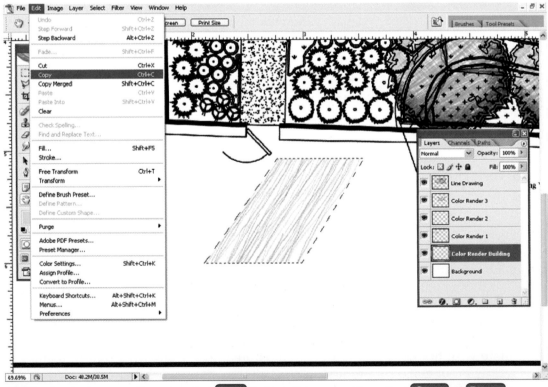

FIGURE 3–97 At the top toolbar tab, select **Edit** and scroll down to either select **Copy** or **Paste**.

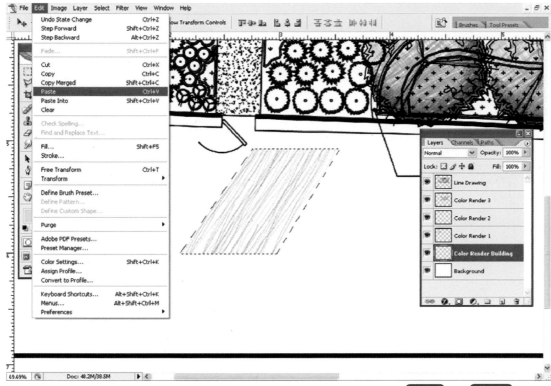

FIGURE 3–98   Create multiple copies of the Parallelogram pattern area using the **Copy** and **Paste** tool.

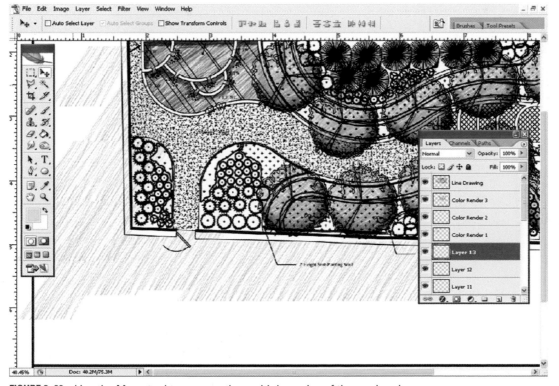

FIGURE 3–99   Use the Move tool to arrange the multiple copies of the rendered area.

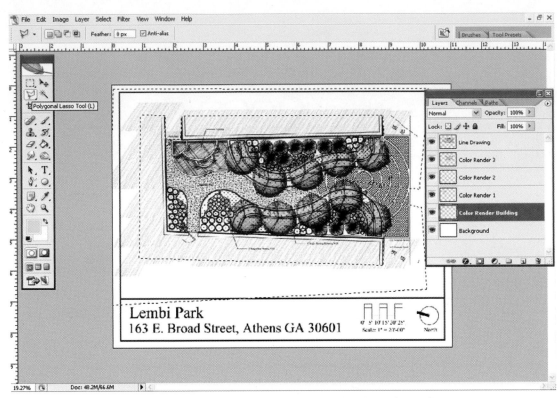

FIGURE 3–100 Merge all the multiple copies of the same color rendered area in one layer.

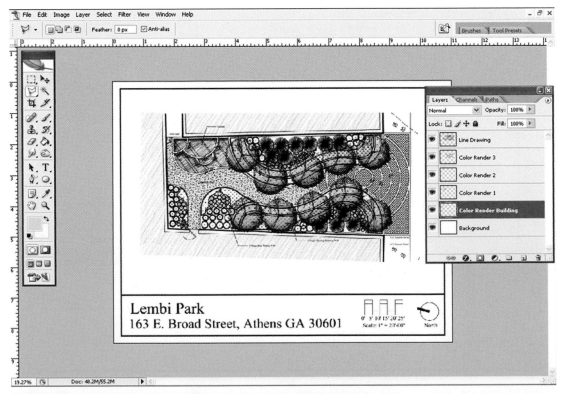

FIGURE 3–101 Use the Marquee and Eraser tool to delete the excess color areas.

Once you render the entire composition, add highlights and shadows to enhance the depth of the drawing. Create a shadow layer and place it below the Black Line layer so that the shadows cover all colors under it (figure 3-102).

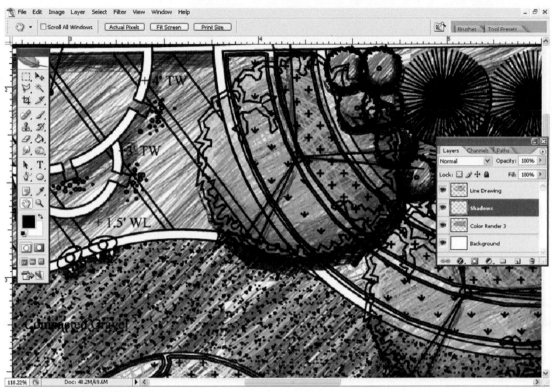

FIGURE 3–102   Create a shadow layer and place it below the Black Line layer in order for the shadows to cover all colors under it.

Merging the layers, saving the image, and plotting

As in the previous blue print, sepia print, and water color exercise, after completing your color pencil rendering, save two copies of the drawing. Save one copy as an active Adobe Photoshop drawing and the second as the finished image. The finished image copy is the one that you need to flatten all layers to create one simple drawing. The active Adobe Photoshop drawing copy you will keep as a backup copy in the event that further revisions are required. **Print/Save** your flattened finished image copy drawing as an Adobe PDF format. The PDF format is printer and electronic transfer friendly, and also reduces the file size (storage capacity) considerably (figure 3-103).

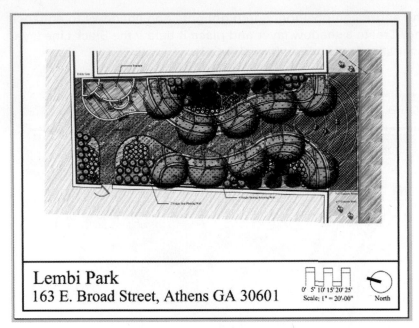

Lembi Park
163 E. Broad Street, Athens GA 30601

0' 5' 10' 15' 20' 25'
Scale: 1" = 20'-00"

North

FIGURE 3–103 The finished PDF file product.

Other Rendering Techniques

Inserting a CAD base drawing into Adobe Photoshop CS2 offers the computer-savvy user the opportunity to explore and experiment with the setting of the brush styles to recreate the feel and texture of any type of hand rendering media. It takes time, exploration, and trial and error to learn what works and what does not. The combinations are as endless as the imagination of the user/artist. This chapter covers only the most frequently used hand rendering media in the professional practice of landscape architecture. Have you tried pastel, charcoal, or even green board chalk? Experiment and combine different tools and media with the Brush Style Editor tool. Write down in your computer rendering journal the detailed steps you took to accomplish the digital rendering effect. Remember, this is the beginning of your digital rendering exploration and discovery, and the future seems even more promising. Have fun!

TERMS

Adobe—Short name for Adobe Photoshop.

Adobe Photoshop—Desktop software application created to manipulate images or photographs on a digital format.

M-Color—An Auto Desk plug-in software available for rendering CAD drawings.

PSD—The acronym for Photoshop Document.

Photoshop Imaging

By Professor Ashley Calabria

CHAPTER OBJECTIVE

This chapter introduces you to Photoshop's image manipulation techniques. By the end of this chapter you should be able to make accurate selections of photographic material, combine them, and create a seamless photorealistic montage.

Introduction to Photoshop for Digital Imagery

Photoshop imaging is a great way to quickly relate to a client what their landscape project will look like through picture quality imagery (figures 4-1 and 4-2). The initial work of selecting images and creating a library of symbols will take a great deal of time, but once it is saved, the plants and materials can be used in subsequent projects. The commands covered in this section are the most commonly used tools for a digital imagery project. In this section, we will create an example image together.

FIGURE 4–1   A digital photo of the site before being designed also referred to as the base image.

FIGURE 4–2   The reconstructed image after being designed.

Author's Notes

- If the cursor turns into crosshairs, you can change it back to the brush size cursor by hitting the Caps Lock key.
- Remember to save images regularly using the default Photoshop or .psd extension.
- Holding down the space bar at any time will allow you to pan around an image.
- Tools with a small black corner have hidden tools underneath them. To access them, click and hold on the tool to show the hidden tools.
- For this exercise, we will keep the background layer locked. To unlock it, double-click on the background layer and give it another name.
- *Always* remember: you must be on the layer that you are trying to manipulate. If you keep selecting or adjusting an item but the effect is not showing, it is probably because you are on another layer. Always check which layer you are on before starting tools.
- The basic steps to create a photo montage are:
    1. Workspace setup: palettes, toolbars, and preferences
    2. Acquire pictures
    3. Setting up the drawing: resize images using dpi and overall size
    4. Selection techniques and moving
    5. Sizing and arranging the source images
    6. Using shadows and effects
    7. Saving and printing: The difference between saving using the .psd, .jpeg, or .pdf file extensions.
- The **base image** is the original image that we will be redesigning. **Source images** are the images used to supply materials for the redesign: planters, plants, people, etc.

The Photoshop Screen

Figure 4-3 illustrates the default Photoshop CS2 screen that will automatically open. Depending on the Photoshop version you have, your screen might look a little different. If you hover over each icon in the toolbar for a few seconds, a yellow tag will pop up with the name of the tool. On top of the screen you will see the default Title Bar showing the file path of where the drawing is located and the name of the drawing. Below the Title Bar is the Menu Bar and then the Options Bar. The main toolbar is typically on the left and the palettes are docked on the right.

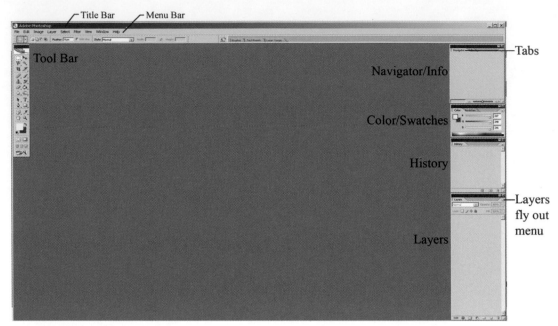

FIGURE 4–3 The Photoshop screen.

Working with Palettes

Set up your workspace with the following seven palettes as seen in figure 4-3: Tools, Navigator, Info, Color, Swatches, History, and Layers. To add toolbars or palettes that might be missing go up to Window and select the toolbars or palettes that you need.

> To merge palettes together, click on the tab of the newly opened palette and drag it to the cluster of palettes that you want it grouped (docked) with.

> To get rid of extra palettes, click and drag onto the workspace area the tab of the palette you do not want. Then you can click on the X to close it.

> To save this workspace so it remembers these settings, go to Window — Workspace — Save Workspace —give it a name— Save.

Toolbar

The most commonly used tools in this chapter are labeled in the toolbar image (figure 4-4). The tools are described more thoroughly throughout this chapter.

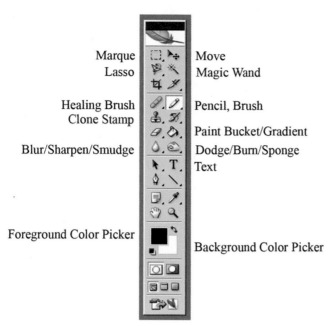

Marque
Lasso

Healing Brush
Clone Stamp

Blur/Sharpen/Smudge

Foreground Color Picker

Move
Magic Wand

Pencil, Brush

Paint Bucket/Gradient
Dodge/Burn/Sponge
Text

Background Color Picker

FIGURE 4–4 The Photoshop toolbar.

Preferences: History States and Zoom

Another important step before starting a project is to set the history states for the History Palette to remember. The History Palette allows you to go back as many steps as needed. It also allows you to go forward as many times as needed as long as another command is not initiated. The default number of commands remembered in the History Palette is 20. To set that to a higher number, go to `Edit` — `Preferences` — `General`, `History States` = 100, then select OK.

At this point you can also select the Zoom with Scroll Wheel option, which allows you to zoom using the scroll wheel on the mouse.

Info and Navigator Palettes

To activate or "bring forward" a tool palette, click on the tab of the hidden palette.

Open the base image by going to `File` and then `Open`. Browse and find the base image to be redesigned.

Select the Info palette tab and then select any tool in the main toolbar. At the bottom of the Info palette, you will see a brief description of the tool you selected. The Navigator palette allows you to zoom or pan quickly. Select the Navigator palette tab to bring it forward and try these simple tasks:

Use the scroll wheel on the mouse to zoom into the image. Holding down the space bar, left-click, and drag to pan around on screen.

Now try using the Navigator palette to do the same things. In the Navigator palette use the slider at the bottom of the palette to zoom in or out. Zoom in close and then move the red box around to pan.

 ## Acquiring Pictures or Image Retrieval

Digital Camera: Resolution

Digital cameras are one of the easiest and most reliable ways to acquire images. The mega pixel capabilities of your camera will affect the print quality at different sizes. Although the display resolution calculated in pixels per inch, or **ppi**, is important, we are going to alter the images depending on what a printer is capable of creating (**dpi**, or dots per inch) and to keep the file size small. We will be creating an image that is approximately 7″ × 10″ using a dpi of 200. To calculate if a camera has enough mega pixels to produce these requirements, use the following calculation:

(height × dpi) × (width × dpi) = mega pixels of camera needed

(7 × 200) × (10 × 200) = 2,800,000 pixels, or a minimum 2.8 mega pixel camera

In general, always take images with the highest mega pixel camera available because pixels can always be removed but not added. Based on the above calculation, a 2.8 mega pixel camera is the minimum needed for this exercise and easy to find; more typical are 7–10 mega pixel cameras, which will work superbly.

Determining the dpi to work with depends on many factors such as the final output size, the printer capabilities, and the ability to manage the file size. A higher ppi creates a larger file size, but currently most printers have the ability to print only at 600 dpi. Even if a picture has a resolution of 1,000 ppi, it will print only with the maximum dpi for that printer. This project uses 200 ppi because for the final size of 7″ × 10″, the image turns out well without creating a huge file size.

When looking for materials to photograph, spend some time in advance to locate items of interest. Go back later to take pictures with the following advice in mind:

1. Take all pictures within a few days for similar weather and seasonal conditions.
2. Take all pictures around the same time of day for lighting conditions.
3. Consider sun angles and orientation for lighting conditions.
4. Plant materials or objects are easier to cut out if they have a solid background.
5. Take pictures as close as possible to the object you want to capture. You can always shrink it down later. Never enlarge an image because Photoshop will add fictitious pixels, which will distort the image.

Some options on adjusting light angles and sun conditions will be discussed throughout the project, but keep in mind that the less adjusting done, the less time your project will take.

Web Images

Web images are generally not recommended unless they come from a site that is specific for image manipulation. Because most Web images are about 72 ppi, they will not work

with our project. Always confirm the quality of the image by opening it in Photoshop and go to `Image` — `Image Size` and check the resolution as well as the size in inches. Remember, you cannot successfully add pixels without shrinking the overall size of the image or you risk losing image quality.

Image Software

Purchasing image software is an easy and effective way to acquire a library of images. Many of these companies cater to landscape architects. The images are made specifically for photo-quality work and generally come in a high resolution. The objects are also already selected out, which is by the far the most time-consuming part of a project like this. Photoshop selection methods are extremely powerful, but due to the intricate shape of natural elements often used in landscape design projects, selecting a good clean cutout can be tedious and time consuming.

Check out http://www.realworldimagery.com. During the time this chapter was written, Realworld Imagery was producing high-quality .tiff images used for advertisers, graphic designers, and other design fields. After you finish this tutorial, go to their website and click on SAMPLES to view and download a variety of high-resolution images for free. They are already selected out, so all you have to do is open it in Photoshop, go to `Select` — `Load Selection` and select channel = Alpha 1. The images can be saved to continue building a symbols library and be used to follow the sequence of commands covered in this chapter.

Scanned Images

A large part of scanning images deals with understanding ppi and dpi, which we have already covered. Keep in mind two general rules related to resolution when scanning in images for this project. First, do not scan in anything under 200 dpi because, regardless of the dpi set for the base image used in this chapter, the print could be grainy looking. And second, it is probably not worth scanning in anything over 300 dpi unless you plan on enlarging the item. Using anything over 300 dpi has the potential to create a large file size.

The next biggest topic to tackle is the variety of extensions available for saving images. Extensions are suffixes used to describe a file type, usually indicating how the file has been saved, in which program, and what the file might contain. For example, .doc files are typically text files saved in Microsoft Word, or .dwg files are AutoCAD drawings. Although we recommend using .jpeg format, a brief description of the most commonly used format extensions follows:

- **.jpeg**: Joint Photographic Experts Group. .jpeg is one of the most commonly used extensions because it creates a relatively small file size, and it can be imported into other programs such as AutoCAD, SketchUp, and InDesign. The only disadvantage is that it groups very similar colored pixels into a single color, which helps make such a small file size but causes what is known as lossy compression. It loses the original color sequence of pixels, but in most cases this is not noticeable. The problem is when the image is saved repeatedly—each time the image is saved, it groups more and more similar colors together. The best way to minimize the effect of lossy compression is to make sure that, after hitting `Save As`, set the Quality under `Image` Options to Max. or 12 in the .jpeg Options Box.

- **.tiff**: Tagged Image File Format. .tiffs are an uncompressed format. The .tiff format is very popular for saving original photos or scans. Because it records the data pixel by pixel, .tiff files do not lose or group information as do .jpeg, but a huge file size is created. You might consider scanning or saving your originals as .tiff but resaving them as .jpeg to work with. That way you always have an excellent-quality original to go back to.

- **.gif** (Graphic Interchange Format), **.bmp** (Bit Mapped Graphics), and **.raw**: These are all considered uncompressed formats and are generally acceptable as extensions for photography and scanning.

- **.psd**: Photoshop Document Extension. This is the default extension when any image has been altered in Photoshop. It allows you to keep your layers, styles, channels, and filters in an uncompressed format so you do not lose any information.

- **.pdf**: Portable Document Format, or the Adobe Acrobat Extension. This is primarily for viewing and e-mailing because saving with this extension creates a relatively small file size.

Setting Up the Drawing: Resize Images Using dpi and Overall Size

Take a look at the image that we will be using for this project (figure 4-1). The first things you will want to do with it are to check its current size and then resize it according to the final output requirements.

FIGURE 4–1 The base image.

Ruler

Control+R turns on the ruler so you can check the actual size of the image.

Right-click in the ruler area to make sure it is set to inches.

Resizing the Image

Go to **Image** — **Image Size** and be sure to check Resample Image, which will get rid of pixels and help reduce the file size. The image size is 10.24″ × 7.68″. Change the width to 10, and the height will adjust proportionately so that the image does not get distorted. Change the resolution to 200, and before hitting OK, check the pixel dimensions at the top of the dialog box which displays the current file size as well as the original file size in parenthesis. You should see the file size drop significantly.

Image Adjustment: Adjusting Brightness, Contrast, Color, or Value

This picture was taken on an overcast day, while many of the source images were taken on sunny days. To adjust the base image so it looks more consistent with the other images, go to **Image** — **Adjustments** — **Levels** and slide the right-hand slider to the left until the input level on the right reads 200 (figure 4-5).

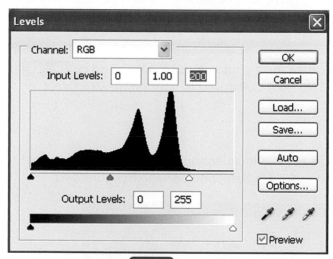

FIGURE 4-5 Adjustment **Levels**.

You can use this for any pictures in the future that need a slight natural adjustment in color. Other common options for image adjustments are under **Image** — **Adjustments**:

The Auto options for levels and color adjust the image by reading pixel information.

Color Balance allows you to adjust several color issues manually.

Brightness/Contrast allows you to adjust brightness and/or contrast manually.

Hue/Saturation allows you to adjust the hue and/or saturation manually.

These commands will give you a preview, so try them all on the base image. Just make sure to hit Cancel or go back in the **History States** to the adjustment level set earlier.

Getting Started

Students are often discouraged with an image of a site in disrepair but to save time and to avoid putting in unnecessary objects, we suggest you go ahead and start designing the site. If there is some clean up work to be done, do it later.

Selecting Out Materials and Moving to the Base Image

For a digital imagery project, one of the most commonly used selection methods will be the Magnetic Lasso located under the Lasso tool. Open up source images 1 and 2 of two different evergreen trees. Resize the image and resolution as in the above exercise (Image — Image Size and set size = 10″ × 7.5″ and resolution = 200), both of which will reduce the file size.

Magnetic Lasso

The Magnetic Lasso reads the difference between pixel colors. Minimize Source 2 image and zoom into Source 1. Activate the Magnetic Lasso hidden under the Lasso tool. Start by selecting along the edge of the evergreen and keep dragging along the edge of it until you loop back around to the beginning point (or double-click to end your selection).

Starting Your Selection Over

End your selection by double-clicking; then start your selection over or go to Select — Deselect, right-click and go to Deselect, or hold Control+D to start completely over.

Adding to Your Selection

To add to your selection hold the Shift key while selecting out the area to add. In the options bar you can also set the option to Add to Selection as seen in figure 4-6.

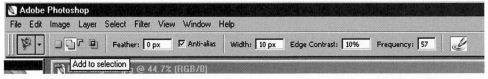

FIGURE 4–6 Setting the options bar for adding or subtracting from a selection in the Options bar.

Deleting Part of Your Selection

To delete part of your selection hold the ALT key while selecting the area to take out. In the options bar you can also set the option to Subtract from Selection.

You can also use these tips with any of the other selection tools to add or delete areas of your selection. Making a good clean initial selection is critical for obtaining a seamless final project and for creating a great resource library of materials.

Freehand Lasso

For more selection detail, zoom into the edge of the tree and use the Freehand Lasso to add to the selection. Add to the selection by using the Shift key or setting the options bar to Add to Selection. This will help with getting a more natural branching structure that was perhaps cut off by the Magnetic Lasso.

Saving and Loading Your Selection

Once your selection is finished, you can save it within this image so that it can be used for other projects without you spending time in making that good clean selection all over again.

Go to **Select** — **Save Selection** — Name and give it a name.

To reload it as a selection later, go to **Select** — **Load Selection** and select the name you gave it. When you save this image as a .psd or Photoshop file, the selection you created and saved will also be saved.

Moving the Source Image to the Base Image

Select the Move tool and then click and drag the tree to the base image. You will see the frame of the base image become highlighted. Release the tree and notice that it has been put on its own layer.

Working with Layers

In the Layers palette, double-click on Layer 1 and rename it Tree 1 (figure 4-7).

FIGURE 4-7   A tree added and the layer name changed.

flipping the Source Image

Because the base image has a slight shadow cast to the right, we will flip this tree so that it appears to have that sun angle. Activate the Tree 1 layer by selecting it in the Layers palette. Go to **Edit**, then **Transform** and select **Flip Horizontal**.

Sizing and Arranging the Source Images

free Transform

Resize the tree by going to **Edit**, then **Free Transform** (Control+T). Scroll out so the **Free Transform** box is in full view. Select one of the corners, and while holding the Shift key, click and drag one of the corners until it looks appropriate in size. Holding the Shift key will keep your tree proportionate. The width and height can also be adjusted by Percentage, located at the top in the options bar. With the **Free Transform** box still active, select inside the box and move the tree to the far right of the back wall on top of the pigeons. Hit Enter for it to accept the changes or hit the ESC escape key to start over.

Drawing Guide Lines Using the Pencil Tool

There are several ways to create an illusion of perspective. This first method is purely by visual perception using **Free Transform** as stated above and resizing the item until it looks appropriate. You can create guide lines using the Pencil tool as an aid by first creating a new layer called "guide lines." Go to the Flyout menu of the Layers palette and select New Layer and name it "guide lines." Select the Pencil tool, and in the options bar at the top of the screen, select the Brush pull-down and set the master diameter to 9. Pick a point at the top of the tree to make a dot. Then go to the corner of the building, and while holding down the Shift key, select a point in the direction of the vanishing point. Holding down the Shift key between two selection points will draw and keep your line straight. Do this from the bottom point of the tree as well (figure 4-8).

When putting in more trees, use the **Free Transform** tool to resize them to fit within the guide lines. Once this is done, you can delete the guide lines layer by dragging it to the trash in the Layers palette or just turn it off by clicking on the eye next to the layer in the Layer palette.

finish the Row of Trees

Source image 2 is of another tree. Open it, resize it, select out the tree (save the selection), and move it next to Tree 1 on the base image. Resize this tree using the **Free Transform**, flip it horizontally for the shadow (right-click inside the **Free Transform** box and select **Flip Horizontal**), and then rename it Tree 2 in the Layers palette (figure 4-9).

FIGURE 4–8   Guide lines drawn on a new layer.

FIGURE 4–9   Another tree added and resized to fit within the guide lines.

Duplicating Layers

One way to duplicate a tree already in the base image is to duplicate the layer it is on. Activate the Tree 1 layer. Select the Flyout menu in the Layers palette and click on

Duplicate (figure 4-10). You can also right-click while on the layer to duplicate, and then select Duplicate Layer .

Name it Tree 3. Using the Move tool, click and drag on the tree you duplicated. You will see a duplicate of it move while leaving the original in the same place. Use Free Transform to scale it down.

FIGURE 4–10 The Flyout menu from the Layers palette.

Duplicate

Another way to make a duplicate is to go to the Tree 2 layer, activate the Move tool, and while holding down the ALT key, click and drag Tree 2. You will see a duplicate of it move, leaving behind the original. Rename this layer Tree 4 and resize it using `Free Transform`.

You can also go to `Edit`, then `Copy` and `Edit`, then `Paste` and use the Move tool for moving the copied tree. Finish the row of trees by duplicating them. Rename the duplicate tree layers Tree 5 and Tree 6. To minimize the repetitious look of using the same two trees, alter their shape and form slightly by using the `Transform` and `Free Transform` commands.

Once all of the trees are in line (figure 4-11), turn off the guide lines layer by clicking on the eye to the left of the guide line layer or get rid of it by dragging the layer to the trash can at the bottom of the Layers palette. Close source image 1 and 2 by saving them as .psd files, which will save your selection within the image.

FIGURE 4–11 Trees added.

Layer Management

Notice that each time you copy a tree or bring in a new tree, it comes in on its own layer. Layers above other layers in the palette will appear in front of those items. To move a layer (or what is visible) above or below other layers, just click and drag the layer, depending on how you want it to be seen. Move your layers around until your trees work in the space. It is important to keep track of what objects are on what layers, so rename sources as soon as you put them on the base image. This gives you more control over manipulating items and organizing how they are seen.

Transform Perspective

Source image 3 is a hedge. Open it, resize it, and zoom into it. The hedge needs some adjustment in color to tone it down. This can be done on the source image itself or after the hedge has been brought over to the base image. This example will adjust the hedge once it is brought over to the base image. Select out a large section of the hedge using the Magnetic Lasso (save the selection) and using the Move tool, drag it onto the base image and over to the far left side of the picture. Resize it using **Free Transform** and then rename the layer Hedge. Go to **Image**, then **Adjustments** and select **Levels**. Adjust the slides until the hedge matches the lighting in the base image (this time the left-hand slide was adjusted to darken the hedge). Use **Free Transform** to resize it to fit along the left wall as seen in figure 4-12. Right-click in the hedge while **Free Transform** is still active and select Perspective or go to **Edit** — **Transform** and select Perspective. This gives you a box similar to the one used for **Free Transform** with grips on the corners. Pick and drag the corners of the perspective box up or down so that the hedge matches the perspective of the base image. Right-click inside the **Free Transform** box again and select Rotate. Hover your cursor outside of the **Transform** box, then pick and drag the hedge to match the angle of the ground. Hit Enter when it is in place. You can also try using **Edit**, **Transform**, and Scale, Skew, or Distort to get it to fit in the space. Save source image 3 of the hedge as a .psd image.

History States

Do not forget that if the hedge gets too distorted from using the **Transform** tools you can hit the ESC key to go back to the original. If you hit Enter and the distortion is accepted, you can click Back in the **History States** palette as far back as you need to go. If you go too far back, you can always click Forward as long as another command has not been started.

Polygon Lasso

The Polygon Lasso is great for selecting items that have straight edges. Source image 4 is of a door and overhang. Open it, resize the image, select out the door and overhang using the Polygon Lasso for the straight edges. Add to your selection using the Magnetic Lasso for the overhang if you need to. Save your selection. Move it to the base image along the left wall next to the shrubs and centered under the two windows above as seen in figure 4-12. Resize the door using the **Free Transform**, then rename the layer Door. Use the Perspective command to put it in better proportion. When it is in place go to the Layers palette, and click and drag the Door layer below the Hedge layer. This will make the hedge look like it is in front of the door.

Save this with the Photoshop (.psd) extension.

Load Selection and Transform Warp

Source image 5 is edging. The edging has already been selected out and saved. Open the image and load this selection by going to **Select**, **Load Selection**, and select 1.

FIGURE 4–12 Hedge added.

Move it to the base image in front of the evergreen trees. Resize it using Free Transform and then rename the layer Edging. Make sure this layer is above the evergreen trees. Use Free Transform to resize it to fit in the space as seen in figure 4-13. Next, we will use Warp to give it a little more bend: go to Edit, Transform and select Warp. You will see a grid appear. Pick and drag any point within this grid area and it will warp the edging to create a smoother curve. If the edging gets too bent out of shape, hit the Escape or Enter key and then go back in the History States. Use Scale and Perspective as well to shift it around until it sits smoothly along the front of the evergreen trees as seen in figure 4-13.

FIGURE 4–13 Transform and a Warp grid applied to edging.

Layer Via Copy

Source image 6 is of gravel and grass. Open it, resize the image, and use the Polygon Lasso to select out a large parcel of both the gravel and grass to fill in the ground plane of the image. Use the Move tool and drag it to the base image and align it so the gravel runs up to the door. Rename this layer Gravel and Grass. Patch some gravel on the right side of the image by activating the Gravel and Grass layer. Use the Rectangular Marquee to select a small rectangular patch of gravel near the spot to patch. Right-click inside the selection and select Layer Via Copy then flip it (using Edit , Transform and Flip Horizontal) and move it into place. Also use this process to patch another area of gravel near the door (figure 4-14). Flip the copy so the texture matches with the original. Drag the Gravel and Grass layer above the copied layers and make sure all of these layers are below the Door, Hedge, Edging, and Tree layers.

FIGURE 4–14 Gravel and grass added.

Merging Layers

The gravel patches are on their own layer and separate from the Gravel and Grass layer. To put them on the Gravel and Grass layer, we will use Merge Down in the Layers palette. Activate the Gravel and Grass layer, which should be above the copied layers, and make sure the gravel patch layers are beneath it in the Layers palette. Go to the Flyout menu of the Layers palette and select Merge Down. This merges the two layers together. Rename the merged layers Gravel and Grass.

Do this with layers that have multiple parts to them. The evergreen trees were not merged because you may still want to shift them forward or backward according to how your other plants work with it.

Layer Via Cut versus Layer Via Copy

Next we will bring the hood of the car in the right corner of the base image into the foreground (figure 4-15). Turn off the Gravel and Grass layer so that you can see the hood of the car. Activate the background layer and use the Magnetic Lasso to select out the hood of the car and along the edges of the image. Right-click in the selection area and select **Layer Via Copy**. Rename the layer Car and move it above the Gravel and Grass layer so that it appears on top of the gravel as seen in figure 4-15. You could use **Layer Via Cut**, but it cuts out the item from the background layer to be put on a separate layer. Use caution when making any changes to the background layer since it is very difficult to fix.

FIGURE 4–15   Car hood brought to the foreground.

Finishing the Image

Try to finish the image on your own using the commands above and the following images: source image 7 of candytuft, source image 8 of an oakleaf hydrangea, source image 9 of flowers for the window boxes, source image 4 of a potted geranium next to the door we already used, and source image 10 of people. If you get stuck, you can follow the steps and the figures below.

- Source image 7 is of candytuft ground cover. Open it, resize it, adjust the levels to tone it down, select it out, and move it as filler in front of the evergreen trees, just slightly to the left (figure 4-16). Resize it using **Free Transform** and then rename the layer Candytuft Left. Duplicate this layer and move this patch to the right. Use Perspective and Distort to fit it in well and adjust the shape (so it does not look like a stamp). Rename the layer Candytuft Right, and place both of these layers at the top of the palette so they are above both the trees and edging layers.

FIGURE 4–16 Candytuft and oakleaf hydrangeas added.

- Source image 8 is an oakleaf hydrangea. Open it, resize it, select it, and move it to go between the 2 patches of candytuft (figure 4-16). Resize it using `Free Transform` and then rename the layer Hydrangea. Arrange the layer so it is above the Edging layer and the evergreen tree but between the candytuft layers. Duplicate this layer, flip it, resize it, and move it over to form a mass of hydrangeas. Make another duplicate for the far right side of the planter and another next to the door. Use `Transform` and `Free Transform` to distort the shapes. Once the mass of hydrangeas on the right is arranged well, use Merge Down in the Layers palette to combine them on one layer. Leave the one by the door on its own layer so it can be moved to fit the space.

- Source image 9 is of some flowers for the window boxes on the brick building. Open it, resize it, select out a batch of flowers, and move it toward the existing window boxes (figure 4-17). Resize the flowers using the `Free Transform`, then rename it Window Box Flowers. To add variety, go back to the source image and select out another patch of flowers to add to the other window boxes. Use `Free Transform` and `Transform` to resize, create perspective, and move them into position. Rename the layers accordingly.

- Source image 4 (the door) has some potted geraniums. Open it, resize it, select out the potted geraniums to the right of the door and move them to the stoop of the door. Resize it using `Free Transform` and rename the layer Potted Geraniums.

- Source image 10 is of people, courtesy of William and Claire. Open it and load the existing selection. Use adjustment levels to lighten them slightly, then move them to the gravel area near the door (figure 4-18). Resize them using `Free Transform`, then rename the layer People.

FIGURE 4–17   Window Box Flowers added.

FIGURE 4–18   People added.

Using Shadows and Effects

Drop Shadow

There are several effective ways to create shadows but they do not consistently work for everything. For instance, a person on a bike has a lot of intricacies that make it more difficult to look realistic than just using a drop shadow or rendering a soft cast.

We will add a simple drop shadow on the hedge, which will help blend the existing bit of shadow along the bottom of it (figure 4-20).

Make the hedge left layer active and go to **Layer**, **Layer Styles** and select **Drop Shadow** (figure 4-19).

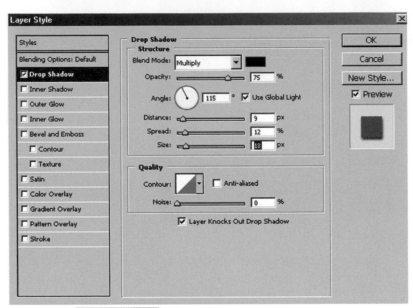

FIGURE 4–19 **Drop Shadow** dialog box.

Change the sun angle to 115° and adjust the distance, size, and spread. Try checking some other options to see how they affect the hedge. Just uncheck them to take the effect off. Check the PREVIEW box so you can see the changes as they are being applied to the hedge. Hit OK when done. Notice that on the Hedge layer a small circle with an F appears on the right. This stands for Effects or what was previously known as FX. To adjust the effects for this layer, double-click on the circle and it will bring you back to the **Layer Styles** set for that layer. Once the sun angle is set, it will remain consistent for all the other layers that use it.

Burn/Dodge Tool

Another way to display a shadow is to simply use the Burn tool. It is located under the Dodge tool. The Burn tool will allow you to darken areas within a layer. Use the Burn tool on the edging and gravel under the candytuft and hydrangeas. This will give the overhanging plants some grounding and depth. Go to the Grass and Gravel layer and activate the Burn tool; in the options bar you can adjust your brush size and style. In this example, the brush is set to soft round 65.

On the Edging layer, brush along the parts of the overhanging hydrangeas and candytuft. You will see the edging darken, casting a slight shadow (figure 4-20). You can use this subtle technique any place that needs a slight adjustment. In the example, the base of each of the trees and hedge was burned, as well as many of the other plants, parts of the wall behind the trees, and in the corner. In contrast to this, the Dodge tool will lighten things up using the same process.

FIGURE 4–20 Shadows created by Drop Shadow and the burn tool.

Intricate Shadows

For larger, more intricate shadows like those of people, we will use the brightness/contrast and opacity to create one and then distort it to cast it onto the ground (figure 4-21). Go to the People layer, right-click, duplicate the layer, and name it People Shadow. Select Image — Adjustments — Brightness/Contrast and slide both down to the left. Edit — Transform — Distort. Click and drag it to the ground. Move the Shadow layer so it is underneath the People layer. Finally, back in the Layers palette, slide the opacity down to 20%.

FIGURE 4–21 More intricate shadows created using balance/contrast.

Blur Tool

To soften some of the edges of the source images, use the Blur tool. Activate the Hedge layer and then select the Blur tool. Run it along the edges of the hedge to soften it. Change the brush size in the Options menu if the brush is too small or large for the areas to touch up.

Vanishing Point

Another great method for creating the illusion of perspective is to use the **Vanishing Point** command. This is a more advanced method and works better with such items as windows or signs. Open source image 11 of the Lembi Park sign, resize the image, and select out the top part of the sign using the rectangular marquee. Save the selection once it is complete. Instead of dragging this selection over go to **Edit** and **Copy**. Back on the base image make a new layer for the sign to be copied too by going to the Flyout menu and selecting New Layer. Name it Sign, make it current, and make sure it is located below the Tree layers. Then go to **Filter** and select **Vanishing Point** and select the four corners of the brick wall where the sign is to go, as seen in figure 4-22.

FIGURE 4–22 **Vanishing Point** dialog box.

A blue box of grids in perspective will be created, demonstrating the effect of the vanishing point. (If the box is red or yellow, try tracing the four corners again with straighter lines.) Hit Control+V to paste the sign. Move it into place and use the **Transform** tool located on the left to resize it while holding the Shift key to keep it proportionate. You will see that as you move your sign onto the grid area it will distort into perspective on its own (figure 4-23). Once the sign is in place, hit OK and it will bring you back to the workspace.

FIGURE 4–23 Sign added.

Clean Up with the Healing Brush Tool

Now that the base image has all the sources added to it, we will clean up the areas that are visible. Starting with the building on the right we will use the Healing Brush tool. It takes a sample of pixels that you select and puts them in a new location but blends them smoothly with the surrounding textures and colored pixels. Make a new layer called Healed Brick and put it just above the background image layer. Activate the healing brush tool, and while on the background image layer, hold down the ALT key and pick a point on the wall close to the vent above the sign, as seen in figure 4-24.

FIGURE 4–24 Area of brick building to clean.

On the Healed Brick layer go back and spot clean up areas of the wall by picking areas close to those that need to be cleaned (figure 4-25). If you start to get a repetition of the wall segment above, go back to the background layer and make another selection by holding down the ALT key. This helps to patch the wall with a variety of samples so it does not look so stamped. Spot clean the worst areas.

FIGURE 4–25   Clone and Healing tools used for cleaning the buildings.

Also clean up the cable on the white wall at the bottom of the left-hand-side building. Try using the Clone Stamp tool located underneath the Healing Brush tool. It works the same in that you take a sample using the ALT key and picking a point while on the background layer, and after going to the Healed layer you can brush on the sample to clean the facade. The Clone Stamp does not try to adjust pixel textures or colors. It is an exact clone of the sample.

To patch the old door next to the one added, go to the background layer and using the rectangular marquee, select out a clear area of wall. Go to **Edit**, **Copy** and then **Edit**, **Paste** or right-click and select **Layer Via Copy**. Move it into place to cover the old door. Make copies of the patch if you need to cover more space. Merge the layers once they are arranged and fully cover the old door. Rename the layer door patch and use Blur to soften the edges. This was also used to patch some brick next to the far left window on the brick building that had a wooden board on it (figure 4-26). At this point, you may want to run a test plot to see how your project is all coming together.

FIGURE 4–26 Buildings cleaned.

Saving and Plotting

.psd

Always save your drawing as a Photoshop drawing or .psd file by going to **File** and **Save As** (figure 4-27).

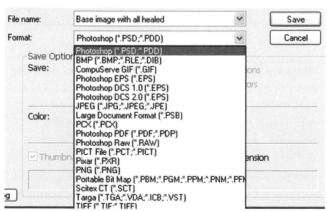

FIGURE 4–27 **Save As** dialog box for saving with the PSD extension.

This will keep your layers separate and allow you to make changes more easily. For printing or e-mailing this image, you would not use this extension because it creates a huge file size and it also gives someone the opportunity to easily change your work.

Flattening the Image

Once the image is finished and saved as a .psd file, go to **Layer** and select **Flatten Image**.
Notice that all your layers have been merged into one background layer. This helps to reduce file size.

Save for Web

The best way to show you how the .jpeg extension reduces file size based on lossy compression is through the **File** and **Save for Web** option (figure 4-28).

FIGURE 4–28 **Save for Web** dialog box.

A closeup preview of the project will show up. At the top of the window select 2-Up. This will show you a comparison between your original and how resolution affects the image quality. Select .jpeg as the file extension. In this box, set the option to low. This allows for maximum compression by combining more colors together into a single color. In the bottom left corner of the window you will see the options you have set as well as the file size. This is best used for e-mailing your image because visual quality is not as noticeable on a monitor. That is also why most images for the Web are saved at 72–100 dpi. But if you were to print this out, you would most likely get a grainy final product. Go back and change the low setting to maximum. Then check the file size again at the bottom left of your screen. You will see the file size has increased significantly. Although the file size is larger then the low-compression option, it is better for saving to print.

Hit **Save**, then designate the location and file name you want to use.

You do not have to use this option every time: once you are comfortable with image sizes and quality you can just go to **File** — **Save As** and save it with the .jpeg or Photoshop .pdf extension option.

TERMS

Base image—The image of the site.

dpi—Dots Per Inch or dpi relates to the density of dots that can be printed in one linear inch on a sheet of paper.

.gif—Graphic Interchange Format, **.bmp**—Bit Mapped Graphics and **.raw**: These are all considered uncompressed formats and are generally acceptable as extensions for photography and scanning.

.jpeg—Joint Photographic Experts Group. .jpeg is one of the most commonly used extensions because it creates a relatively small file size, and it can be imported into other programs such as AutoCAD, SketchUp, and InDesign. The only disadvantage is that it groups very similar colored pixels into a single color, which helps make such a small file size but causes what is known as lossy compression. It loses the original color sequence of pixels, but in most cases this is not noticeable. The problem is when the image is saved repeatedly—each time the image is saved, it groups more and more similar colors together. The best way to minimize the effect of lossy compression is to make sure that, after hitting **Save As**, you set the Quality under **Image** Options to Max. or 12 in the .jpeg Options Box.

.pdf—Portable Document Format or the Adobe Acrobat Extension. This is primarily for viewing and e-mailing because saving with this extension creates a relatively small file size.

ppi—Pixels Per Inch or ppi relates to the number of pixel boxes that occur in one horizontal or vertical inch of your computer display.

.psd—Photoshop Document Extension. This is the default extension when any image has been altered in Photoshop. It allows you to keep your layers, styles, channels and filters in an uncompressed format so you do not lose any information.

Source images—The images used for redesigning the site.

.tiff—Tagged Image File Format. .tiffs are an uncompressed format. The .tiff format is very popular for saving original photos or scans. Because it records the data pixel by pixel, .tiff files do not lose or group information as do .jpeg, but a huge file size is created. You might consider scanning or saving your originals as .tiff but resaving them as .jpeg to work with. That way you always have an excellent-quality original to go back to.

InDesign

By Professor Ashley Calabria

CHAPTER OBJECTIVE

This chapter introduces you to InDesign's desktop publishing capabilities. By the end of this chapter you should be able to create a booklet or poster including text, images, and master page elements.

Introduction to Adobe InDesign

InDesign is Adobe's award-winning desktop publishing application. It allows the designer more control over image and text manipulation for page layout of documents, brochures, or sheets. Current uses in landscape architecture are for portfolios and/or sheet layout of projects that contain a variety of images from different resources. The setup for booklet or sheet is very similar, but the booklet format tends to require more detail and repetition. This chapter will follow the processes through booklet format, extracting information for sheet development along the way.

Author's Notes

- Tools with a small black corner have hidden tools underneath them. To access them you will need to click and hold on the tool to show the hidden tools.

- As with many computer programs, there are a variety of ways to activate a command or manipulate an item. As you get more comfortable with InDesign you will find alternative ways to access commands or perform tasks. Use what works for you.

The InDesign Screen

When you first open InDesign, select Close. We will set up a new document after look-ing at the workspace. Figure 5-1 illustrates the default InDesign CS2 screen that will open. Depending on the version you have, your screen might look a little different. If you hover the cursor over each icon for a few seconds, a yellow tag will pop up with the name of the tool. To add toolbars or palettes that might be missing, go to **Window** and select the toolbars or palettes that you need.

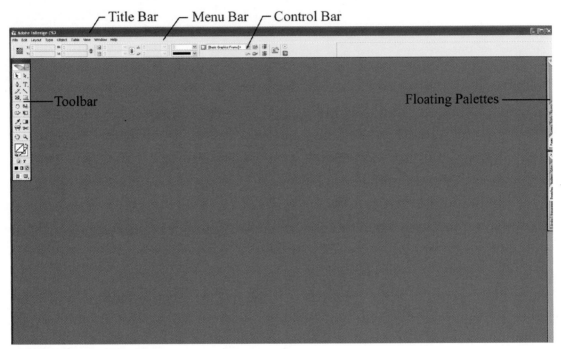

FIGURE 5–1   The InDesign screen.

On top of the screen you will see the default Title Bar showing the program, name of the drawing, and zoom factor. Below the Title Bar is the Menu Bar. The Control Bar changes to provide quick options for select items on the page. The most commonly used tools are along the left toolbar. When a document is open, you will see the Status Bar at the bottom of the screen. On the left side of the Status Bar is the document page number, zoom factor, and filename. On the right side of the workspace are the Floating Palettes.

Working with Floating Palettes

The Floating Palettes are similar to palettes in Photoshop but usually remain docked to the side. To activate or bring out a palette from being docked, click on one of the tabs. To activate a tab, click on it. To remove tabs, drag the tab into the workspace and close it by clicking on the X button also known as the close box. To bring a tab into a Floating Palette, select the tab name from the **Window** menu and then drag it to the Floating

Palette you want it to be docked in. The tabs in the Floating Palettes will vary but the ones that will be most used in this chapter are Pages, Links, Color, Swatches, and Stroke. These can be arranged in any order in any Floating Palette. You can also keep transparency and gradient tabs in a palette, although we will not use them much. To save this workspace so it remembers these settings, go to `Window` — `Workspace` — `Save Workspace`. Give the workspace a name and select `Save`.

Toolbars

The most commonly used tools in this chapter are labeled in the toolbar image in figure 5-2. The tools are described more thoroughly throughout this chapter. Since the Control Bar varies based on the tool or the item selected, it will be described and labeled throughout this chapter as it is used.

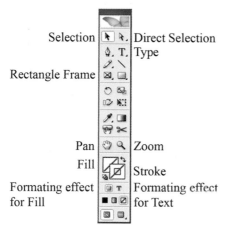

FIGURE 5–2 The InDesign toolbar.

 ## Basic Document Setup

For this chapter, we will create a small five-page graphics booklet for the hypothetical site Lembi Park. It will be of standard letter size, double sided with binding in the middle. If you have a document open in InDesign, close it so we can set up some preferences before beginning.

Setting Up Preferences

Setting up preferences allows you to adjust a variety of features that relate to the display of the document and desktop. We will adjust the display performance setting and the units and increments going to `Edit` and selecting `Preferences` and then `Units and Increments` (figure 5-3).

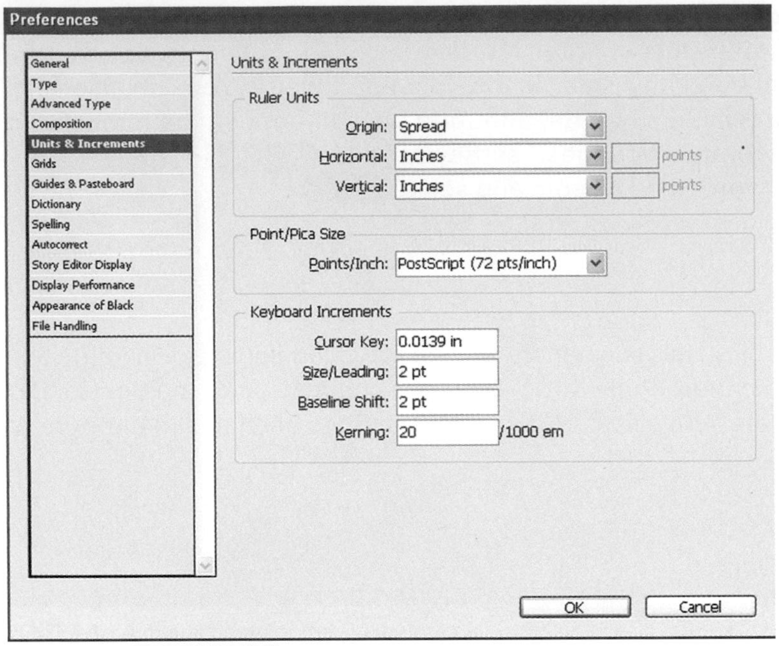

FIGURE 5–3 Preferences dialog box for adjusting units and increments.

Under Ruler Units on the right, there are three settings that need to be adjusted: the Origin, the Horizontal Ruler Unit, and the Vertical Ruler Unit. The Origin describes how the document will be measured. Selecting Spread will start the ruler of the document in the upper left-hand corner at 0″ and read all the way through to the right. So if the document is double sided and 8½″ × 11″, the ruler will read 0″ all the way to 17″ across the top of the screen. Selecting Page will set a ruler that starts at 0″ in the upper left-hand corner of the document until it reaches the next sheet, where the numbering will start again at 0″. So in our example, the ruler would read 0″ to 8½″ and then start over from 0″ to 8½″ on the next page. Selecting Spine will set the ruler to start at 0″ until it reaches the spine of a multipage spread and then will start over with 0″ for the remaining sheets in the spread. Since our document is only a double spread with same-size pages, keep the Origin to read Page. For a poster, you may want to set the ruler to read Spread.

Set the Horizontal and Vertical increments to Inches.

In the left-hand column, select Display Performance. Set the default view to High Quality and check the Preserve Object-Level Display Settings. Also, set the Adjust View Settings to High Quality. This allows you a more accurate display of the images and resolution. So if your images are coming in grainy even after setting the preferences here to High Quality, you may need to go back and get a better quality scan or save the image with a higher resolution. For larger documents, this would slow down the program considerably, but since our document is small, even high-quality images should be fine.

Setting Up Folder Options for Image Links

Setting up folder options for images allows you to directly access, alter, and relink your images between InDesign and Photoshop. To do this, minimize InDesign and go to

My Computer; from the Menu Bar select **Tools** — **Folder Options** — **File Types** and scroll down until you find .jpeg (assuming the images you will use are .jpeg images). Select the Change button and scroll through the programs to find Photoshop CS2 (figure 5-4).

FIGURE 5–4 Setting up **Folder Options**.

Starting a Document

Go to **File** and select **New**, then **Document**. Figure 5-5 shows the dialog box for setting up a new document.

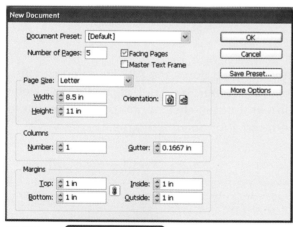

FIGURE 5–5 **New Document** dialog box.

Set the number of pages to 5, and check facing pages. Keep the Page size as Letter and the Orientation as Portrait. The margin settings are used as a guide; you can still print beyond your set margins. For a single sheet, just set the pages to 1 and page size to a custom size for the poster you plan to use. For custom document setup, you can save your **New Document** parameters by clicking on the Save Preset button on the right and giving the document a name. For this tutorial your screen should look similar to that in figure 5-6.

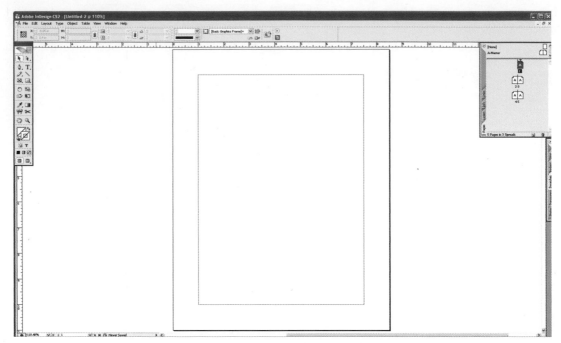

FIGURE 5–6   InDesign screen with an open document and five pages in the Pages Palette.

You will also notice that, in the Pages Palette, page 1 is a right-sided single nonfacing page of the booklet. This is because in standard desktop publishing for books, page 1 is always a right-sided page. To get around this for other documents that might need a facing page as page 1, you can start your work on page 2.

Zoom, Pan, and Pages Palette

Zoom in and out by using the Zoom tool while holding down the ALT key. You can also select a variety of zoom options under View in the Menu Bar. Look at the Pages Palette and you will see all the pages in your document. Double-click on the page you want to go to. The Pan tool will allow you to move around your document area.

Text Tool

Creating a Textbox

The Text tool allows you to create boxes for inputting text. Activate the Text tool and click and drag to create a box on the first page of your document. The blue box is called the Frame Edge and will not print unless the textbox is assigned a stroke. Use the Select tool to select the box so it is active or highlighted. Using the Select tool allows changes to be made to the box itself. So to erase the box, use the Select tool and hit the Delete key. To enlarge or shrink the box, click and drag one of the corners or sides of the box and resize it. But using the Selection tool allows for more specific manipulations as well. With the textbox selected the Control Bar at the top of the screen is as seen in figure 5-7.

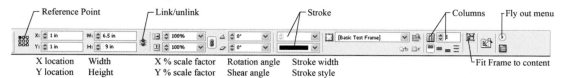

X location   Width   X % scale factor   Rotation angle   Stroke width   Fit Frame to content
Y location   Height   Y % scale factor   Shear angle   Stroke style

FIGURE 5-7   The Control Bar with a textbox selected.

The Reference Point identifies the point at which the selected item (in this case, a textbox) is located within the document based on the X and Y distances. In figure 5-7 the upper left-hand corner of the selected textbox is 1″ over and 1″ down from the edge of the sheet. Remember, the margins you see are only guidelines: you can still print beyond them so the X and Y distance relate to the edge of the sheet of the paper, not the margins.

The X and Y distances identify or determine the location of the selected item based on an identified reference point. You can enter in a specific location for items to be placed based on different reference points and X and Y locations.

W and H stand for width and height of the selected item. Keeping the link sign will adjust the width and height proportionately to the original size specifications. To unlink them, so that the width and height of the box can be manipulated or given different sizes, click on the link symbol and you will see a broken link appear. Set the reference point to the upper left corner as X = 1, Y = 1, W = 6.5, and H = 9 as seen in figure 5-7.

Similarly, the X% scale factor and Y% scale factor will enlarge or shrink the box by percentage based on the original size. A 100% will keep the box the same size, a smaller number will shrink the box, and a larger number will enlarge it. Linking will keep them proportionate, or you can unlink the percentages by clicking on the link symbol.

Rotation and shear angle will allow the box to be rotated or skewed.

Stroke refers to putting a border around the textbox, which will be covered later.

Columns allows you to create columns of text within the textbox.

Selecting Fit Frame to Content adjusts the frame edges of the box to surround any text in the box.

The Flyout menu lists some options for rotating and flipping the selected item.

Stroke

Stroke refers to putting a border on a selected box, either a textbox or picture box. Select the Stroke tool in the toolbar so it sits in front of the Fill tool. With the textbox still selected, go to the Stroke Size drop-down menu in the Control Bar and select 3 pt. You can also select a Stroke Style in the drop-down menu. To give it a different color, open the color tab in the Floating Palette to the right of the workspace and from the Flyout menu select the RGB color mode as seen in figure 5-8.

At this point you can enter in the RGB color numbers for a specific color or slide the bars on the color sliders until you find a color you like. This example will use R = 221, G = 71, B = 44 for creating a red color. Once a color has been found, go back to the Flyout menu of the color tab and select Add to Swatches. When you click on the swatches tab, you should see your color listed there.

Use the upper left corner of the Reference Point and change the X and Y location to 0.75. Also change the width to 7 and the height to 9.5. This will set the large red frame just outside the margins of the paper but still within printing range.

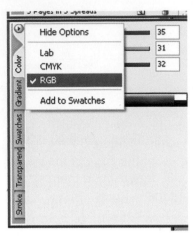

FIGURE 5–8 Selecting a color to add to the color palette that will be used as a stroke color for the textbox.

Fill

With the textbox selected go to Object—Content and select Unassigned from the Menu Bar. If you try to add a new textbox on top of the existing one, it will try to add type to the existing box instead of allowing you to create a new textbox. Now use the Type tool to create another textbox on top of the previous one, select it, and type in a Width = 6.5 and Height = 3. Set the reference point to the upper left corner of the box and set the X = 1 and Y = 1.

Select the Fill tool in the toolbar so it sits in front of the Stroke tool. To give it a different color, open the color tab in the Floating Palette to the right of the workspace, and from the Flyout menu, select the RGB color mode, as we did with the Stroke. At this point you can enter in the RGB color numbers for a specific color or slide the bars on the color sliders until you find a color you like. This example will use R = 237, G = 138, B = 83. As you adjust the RGB color you should see the textbox fill with that color. Add it to the Swatches by going to the Flyout menu and select Add to Swatches.

Adding Text

Select the Text tool again and pick inside the filled textbox. A cursor will appear in the upper left corner of the textbox. At the top you will notice that the Control Bar has changed.

On the far left of the Control Bar, select the A, which stands for the Character Control Bar (figure 5-9). This information deals primarily with how the text appears.

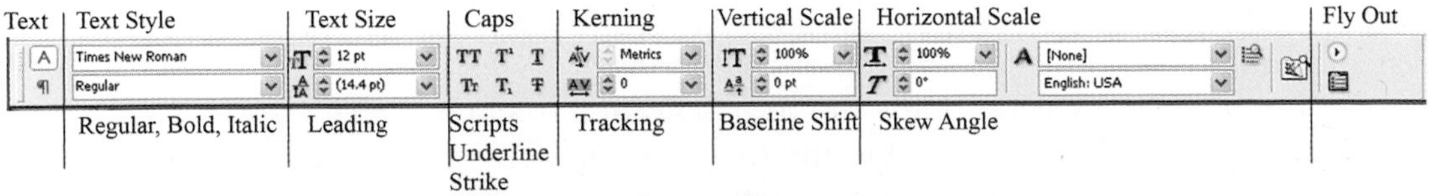

Text	Text Style		Text Size		Caps		Kerning		Vertical Scale		Horizontal Scale					Fly Out
A	Times New Roman	▾	IT ⬍ 12 pt	▾	TT T¹ T	A⁄V ⬍ Metrics	▾	IT ⬍ 100%	▾	T ⬍ 100%	▾	A [None]	▾			▸
¶	Regular	▾	A̬ ⬍ (14.4 pt)	▾	Tr T, T	A⁄V ⬍ 0	▾	A⬍ ⬍ 0 pt		T ⬍ 0°		English: USA	▾			▤
	Regular, Bold, Italic		Leading		Scripts Underline Strike		Tracking		Baseline Shift		Skew Angle					

FIGURE 5–9 The Control Bar for the Text tool Character controls.

On the far left of the Control Bar, select ¶ for the Paragraph Control Bar (figure 5-10). This information deals primarily with how the entire paragraph appears.

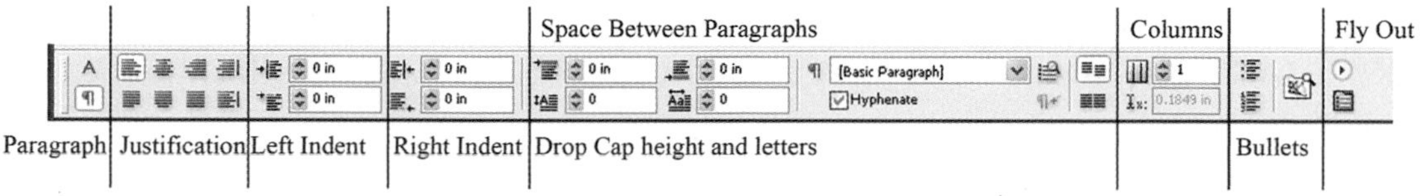

					Space Between Paragraphs						Columns				Fly Out

| A | | | | | 0 in | | 0 in | | 0 in | | 0 in | [Basic Paragraph] | | | 1 | | | |
| ¶ | | | | | 0 in | | 0 in | | 0 | | 0 | ☑ Hyphenate | | | 0.1849 in | | | |

Paragraph | Justification | Left Indent | Right Indent | Drop Cap height and letters | | Bullets |

FIGURE 5–10 The Control Bar for the Paragraph controls.

Select A to access the Character Controls and set the Font Style to Magneto with a size of 72 and type Lembi Park. Highlight it and set the Tracking to −10 and the Vertical Scale to 125%. With the text style highlighted, click on the Fill tool in the toolbar and change the color of the text in the Swatches Tab to Paper. This will allow the text to print whatever color the paper is. To add a Stroke around the text, keep it highlighted and set the Stroke tool in front of the Fill tool, then select a color in the swatches or color palette. The text here has no Stroke.

Create another textbox near the bottom of the document and set your text style to Century Gothic with a text size of 48. Type in "A Graphic Catalog." Select the text and set the tracking to −30 and the Vertical Scale to 110%. Change the fill color of it to the redder color that was added to the swatches. Now use the Selection tool and select the box. Right-click on the box and select Fitting—Fit Frame to Content. Move the box to overlap Lembi Park and along the right-hand margin, as seen in figure 5-11.

FIGURE 5–11 Overlapping textboxes.

Create another textbox near the bottom of the document and set the Width = 6.5, Height = 5.75 and fill it with the orange color. Set the X = 1 and the Y = 4.25 based on an upper left corner reference point. Set Text Style to Century Gothic and Bold,

Text Size to 14, and Tracking to −50. Also select All CAPS or set your caps lock on. Using the Text tool, select inside the textbox and hit Enter to skip down a line, then type AUTOCAD—two spaces—period—two spaces—SKETCHUP—two spaces—period— two spaces—PHOTOSHOP RENDERING—two spaces—period—two spaces—PHOTO-SHOP IMAGERY. Under the Paragraph Control, select the Center Justification. Leave the text black.

Viewing Your Work

The text frames and guidelines often interfere with getting an accurate view of your work. To see your page without those items, go to View, Grids and Guides and select Hide Guides or select the Control key and semicolon, which will remove the margins or any guidelines. You can also remove frame edges by going to View then Hide Frame Edges or select Control+H. To turn them back on select Control+; and Control+H again. Pressing the W key will hide grids, guides, and frame edges all at once. Figure 5-12 illustrates the document without frame edges or margin guides.

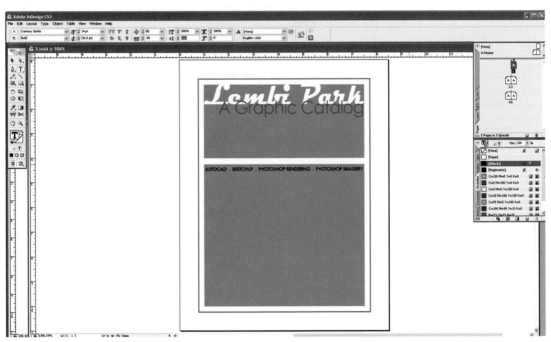

FIGURE 5–12 Viewing the document with no frame edges or guidelines.

Lines

We will use the Line tool to create some decorative lines across the document. Select the Line tool and select a point on the left margin in the center of the gap between the two textboxes. Holding down the Shift key will make a straight line. Drag your line to the right margin and release the mouse. Set your Stroke Type to thin-thick-thin and a Stroke Size to 5 pt. Select the Stroke tool in the toolbar and select the red color from the swatches. Draw two more solid lines with a thickness of 4 and set the color to paper. Draw one above the gap near the bottom of the top textbox and the other line below the gap but above the text in the bottom textbox (figure 5-13).

FIGURE 5–13 Finished cover of the document.

Master Pages

Setting up **Master Pages** is like designing a background of items that occur on multiple pages within a document. When items like textboxes, lines, or picture boxes are created on a Master Page, they are called Master Objects. Once the Master Page is designed with Master Objects, it can be applied to specific pages within your document. Once it is applied to pages in the document, the objects of the Master Page are not easily manipulated but can be changed all at once from the Master Page in the Pages Palette. This allows for quick alterations in the layout but keeps your document consistent without having to update changes for every page individually. You can have multiple Master Page layouts for different chapters or for left- and right-handed pages.

The Master Page information is located in the Pages Palette (figure 5-14).

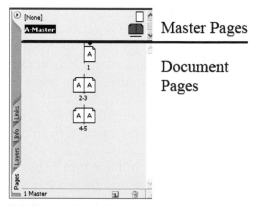

FIGURE 5–14 Pages Palette.

Creating a Master Page

Although you can design a Master Page in the Master Page section, the layout that we created on page 1 will actually be our Master Page for the document. Using the Selection tool, make a window around the entire page. Right-click inside the page and select Cut. In the Pages Palette, double-click on the sheets of paper to the right of where it says A-Master. Notice that a double-sided blank document will appear (figure 5-15). Right-click on the right-hand page and select Paste In Place. With the items still selected, hold down Alt to make a duplicate of the entire layout and hold down the Shift key to force it to move horizontally and slide the duplicate to the left-hand page. You can use the X and Y location of 0.75 and a reference point of the upper left-hand corner to place the entire layout accurately.

FIGURE 5–15 Doubled-sided Master Pages: A-Master.

You can now work with Master Objects in the Master Pages, you can alter existing objects in the Master Pages, and you can create Master Objects in the Master Pages section. Create new Master Pages by clicking on the Flyout menu of the Pages Palette and select New Master Page. It will create a new set of double-sided pages in the Master Pages section of the Pages Palette, called B-Master. If you want many of the Master Objects from A-Master to appear on other Master Page setups, click on the Flyout menu of the Pages Palette and select Duplicate Master Spread A-Master.

To Apply Master Pages to different document pages, just click and drag the Master Page to the document page in the pages palette. To see how this works better, from the Master Pages, click and drag the None Page to document pages 2 and then to 3. Now double-click on page 1 in the Pages Palette. Scroll down and you will see that pages 2 and 3 are blank because they are following the None Master Page and pages 4 and 5 have the A-Master page applied to them. You can override any Master Page on a document page in the same way you overwrote the A-Master with the None Page. Go ahead and replace the A-Master on the document pages 2 and 3, then double-click on page 2.

Overriding Master Objects

You will notice that when you try to select any of the Master Objects on page 2 of the document, nothing gets selected. Override Master Objects to alter them on a page by going to the Flyout menu of the Pages Palette and selecting Override All Master Page Items. Now you can double-click on the textbox with the subheading AUTOCAD. SKETCHUP.PHOTOSHOP RENDERING.PHOTOSHOP IMAGERY. Highlight all of it except the word AUTOCAD. Make sure the Fill tool is in front of the Stroke tool in the toolbar. In Color Tab select a light grey color. In this example 20% black was used by entering 20 in the slider bar with T (for transparency) next to it. Click on the Flyout menu and add this to the swatches. The highlighted text should be grey.

Picture Boxes

Creating Picture Boxes

The Rectangle Frame tool allows you to create boxes for inputting pictures. Activate the Rectangle Frame tool and click and drag to create a box on the second page of your document in the middle section of the orange textbox. Similar to a textbox, the blue box is called the Frame Edge and will not print unless the rectangle frame box is assigned a stroke. The X through it distinguishes it from a textbox. Use the Select tool to select the box so it is activated or highlighted. Using the Select tool allows changes to be made to the box itself. So to erase the box, use the Select tool and hit the Delete key. To enlarge or shrink the box, click and drag on one of the corners or sides of the box and resize it. Using the Selection tool allows for more specific manipulations as well. With the rectangle frame box selected take a look at the Control Bar at the top of the screen as seen in figure 5-16.

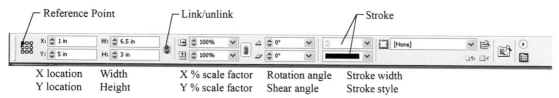

FIGURE 5–16 The Control Bar for the Rectangle Frame tool.

The Control Bar for the Rectangle Frame tool is similar to the Control Bar for the Text tool. Applying a Stroke and Fill are the same as well. Using an upper left corner reference point set the X = 1, Y = 5, W = 6.5, and H = 4 in the Control Bar.

Placing Pictures

Before placing pictures in your document make sure that all the images you plan on using are resized and also saved as JPEG files. I usually resize images in Photoshop to whatever paper size I am using in InDesign. So if the image is originally 24″ × 36″ and I want to insert it in a document that is 8½″ × 11″, then I will resize it to 8½″ wide with the resample image checked so that Photoshop knows to get rid of the excess pixels. This will help keep the overall document folder file size smaller while still giving you good-quality prints.

To place a picture in this box, go to **File**—Place or hit Control+D. This will open a browser for you to search your files and find the AutoCAD project that should also be saved as a .jpeg. Hit Open. You will notice that the image is larger than the box. Right-click inside the image area and select Fitting, Fit Content Proportionately. This will shrink the image so that it fits proportionately within the size of the box that has been created (figure 5-17). You will see that it will not fit perfectly because the proportions of the image and box are different.

FIGURE 5–17 A document with text, lines, and an image.

Direct Select

Use the Direct Select tool in the toolbar and select the image. The Direct Select tool is different from the Select tool in that it selects the image inside the box to be manipulated. So if you want to delete the image and not the box, use the Direct Select tool, select the image, and hit Delete. Also notice the Control Bar at the top of the screen with the Direct Select tool. It is the same as the Select tool. Because the image was scaled down to fit inside the box we created, you will see that the horizontal and vertical scale factor percentage has been reduced. Right-click inside the image again and select Fitting, Center Content. You can also right-click and select Fitting, Fit Frame to Content. This forces the Rectangle Frame to run along the edge of the image, so if a Stroke is applied, it would run along the edge of the image, not the original box created for the image.

Now finish pages 3, 4, and 5, starting with overriding the Master Page Objects. In this example, the two larger textboxes were selected and the Fill was set to None. The white lines were changed to orange and the Lembi Park title was selected and given a Stroke of black (figure 5-18).

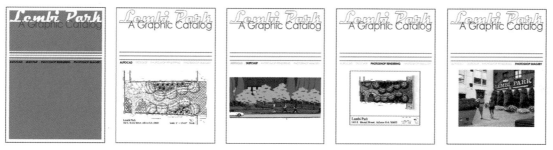

FIGURE 5–18 The document with images and changes made to the Master Page.

Links and Images

Finding images within your document, making changes to images, and getting information about images are done from the Links tab. Click on the Links tab and on an image in the document. The image filename will be highlighted in the Links tab. To find an image from the Links tab list, select the filename and then click on Go to Link at the bottom of the palette (figure 5-19).

FIGURE 5–19 Go to Link from the Links tab.

The image will appear on the screen. Double-click on a filename in the Links list to find out more information about it. To alter an image that has already been inserted, click on the image and on the pencil at the bottom of the Links palette. This is the **Edit** Original option and will automatically open your .jpeg image in Photoshop. This is why we set our folder options earlier to open .jpegs with Adobe Photoshop. Once in Photoshop, you can make any changes and resave the image. It will automatically update the image in InDesign (figure 5-20), keeping all of the original settings that you used in the Control Bar.

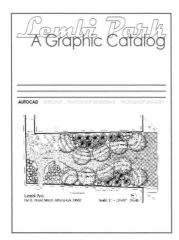 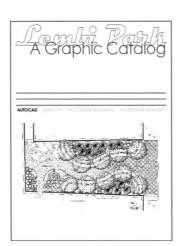

FIGURE 5–20 The original AutoCAD image (with title, address, and scale) was altered in Photoshop and resaved, updating the image automatically in InDesign.

Saving

Package

Adobe InDesign does not automatically embed images into a document. This helps in reducing the file size. You can embed them from the Links tab by selecting the image and then selecting Embed File from the Flyout menu. If you do not embed them or package them, then you will get a ghost resolution of 72 dpi in place of your original high-quality image that did not print well. To harness all of the fonts and images that are in a document, go to **File** — **Package** — Continue. Continue until you get to a browse for a folder location. Select a folder location and name, then hit **Package**. This will save your document, any fonts you used, and your images all under one folder. That way, InDesign knows right where to find them. Remember that if you make any changes to images, choose the images from that folder or else InDesign might not recognize the link.

author's note

- If your images still come in grainy, go to View, Display Performance and set it to High Quality Display. This will give the most accurate view of how your image quality will print.

- For automatic page numbers, go to your Master Page and create a textbox in the location and with the font style and size you want. Go to Type—Insert Special Character—Automatic Page Number. This will appear as a capital A in your textbox, but in your document it will display the page numbers correctly.

- To move pages around, just click and drag them within the Pages Tab.

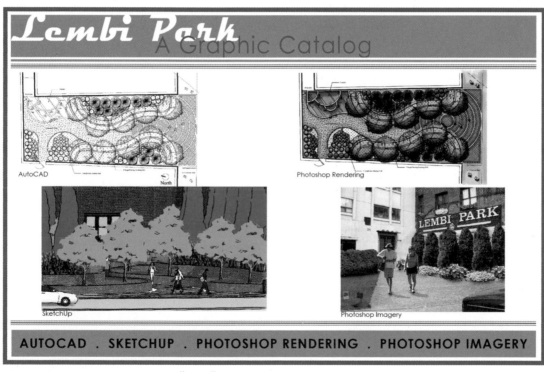

FIGURE 5–21 The document as a 36″ × 24″ poster.

TERMS

Master Pages—Setting up Master Pages is like designing a background of items that occur on multiple pages within a document.

Package—Adobe InDesign does not automatically embed images into a document. This helps in reducing file size. The Package command allows you harness all of the fonts, text and images that are in a document so they are under one folder.

Program Interchange and Student Project Examples

By Professors Ashley Calabria and Jose R. Buitrago

CHAPTER OBJECTIVE

This chapter introduces the reader to the program interchange process used for the graphic development of a professional landscape architecture project. By the end of this chapter you should be able to exchange a drawing between the various computer programs highlighted in this book. The end of this chapter showcases a visual inventory of students' work that demonstrates the use of a mixture of mediums and programs.

 ## Program Interchange

Today's landscape architecture professional practice benefits greatly from the exchange between traditional hand graphics and digital graphics. Depending on the user's computer skill level, proficiency, time frame, and budget, the hand or digital format may be favored for a specific phase of the design process. Traditionally, hand graphics are used in the early development phase of the landscape architecture project because of its flexible nature and the artistic creativity associated with it. During this phase, artistic representations such as bubble diagrams, conceptual plans, section-elevations, axons, and perspectives are common. These artistic representations are rendered via a mixture of mediums that best fit the project and client's taste. Computer graphics have traditionally been used at the end phase of the project for construction documentation and master plan development because of the ease of revisions, cost, and the ability to print multiple reproductions. Technological advances in digital software have changed the traditional place and use of computer graphics in the landscape architecture project.

The computer programs used in this book have been selected based on their popularity in the field of landscape architecture and have been supported by the survey information described in the preface. But the entire series of graphics created for a project rarely are limited to a single application or medium. Although program interchange has been mentioned in various chapters of this book, tying it all together can still be confusing. It takes time and practice to determine at which point a drawing gets sketched by hand and/or drafted in AutoCAD, rendered with markers and/or rendered in Photoshop, or sketched by hand and/or modeled in SketchUp.

Maneuvering through the various computer programs depends largely on understanding file extensions and compatibility. Extensions are suffixes used to describe a file type, usually indicating how the file has been saved, in which program, and what the file might contain (see Chapter 4 for more detailed information on file extensions when scanning). At publication, the programs used in this book have different file extensions that may or may not be readable in other programs. For instance, the most common way to exchange program information for completing a project might use a digital file for a base plan—sketches or preliminary drawings done by hand and scanned into AutoCAD for drafting. Once approved, the file can be converted for rendering in Photoshop, printed out for some additional hand rendering techniques, and most likely imported from AutoCAD into SketchUp for quick 3D modeling to be ultimately used for tracing over by hand. The final series of drawings could be brought together in InDesign for a document or boards for presentation. This is why it is important to understand program interchange and file extensions.

AutoCAD to Photoshop

AutoCAD 2007 does not currently have a direct saving method for exchanging DWG files into file formats that Photoshop can read. There are two easy ways to address this exchange, both of which should be performed in layout (not model space). One is to go to **File** and then Export and set File of type = .eps, or Encapsulated Post Script, so as to be able to bring the file into Photoshop. Then select a location, a name and save it. In Photoshop, go to **File** and **Open**, find your file and select **Open**. A Rasterizing Generic .eps format dialog box appears, where you can adjust the width, height, resolution, and color mode. Depending on the project, I usually select 36″ wide, allow the height to adjust proportionately, adjust a resolution of 300, and select RGB color mode. Checking anti-alias helps smooth the lines in transition from vector to raster format and constrain proportions keeps the width and height linked so the project dimensions do not get distorted.

Another way to transfer files between AutoCAD and Photoshop is to convert the file to a .pdf. The easiest way to work with .pdf is to set .pdf as the printer in AutoCAD layout or to go to **File** — **Plot** and set .dwg to .pdf or Adobe PDF as the printer. Set your other plotting parameters and then select **Plot** and choose a location, name and save it. In Photoshop, go to **File** and **Open**, then select .pdf file. In the dialog box, select Crop to Media Box (which keeps the same paper size), set the resolution to 300 and color mode to RGB unless you are using other standards. Checking anti-alias keeps the lines smoother, and I usually leave the drawing as 8 bit.

AutoCAD to SketchUp

Although you can import .jpeg, .tiff, and .bmp files, using the .dwg extension is the most interactive way, allowing SketchUp to read existing lines from AutoCAD to help create the surfaces needed for quick 3D modeling. Keep in mind that at the time this book was written, Google SketchUp 6 will not read AutoCAD 2007; so, when saving the drawing, select AutoCAD 2004 or an earlier version. To open it in SketchUp, go to **File** — **Import**. For Files of type, select .dwg. Once the Import Results dialog box has finished, you can close it and start tracing your work to create surfaces or insert components. You can check to verify or change the scale of the drawing by using the Tape Measure across a known distance and checking the value control box. If it does not read the distance correctly, you can type it in at this time and then select the Resize the Entire Model option in the box.

SketchUp to Photoshop

While working in SketchUp, create the scenes that will be printed or used. This allows for more detailed materials or components without bogging down the computer and it also helps keep consistent with selected views. Zoom, orbit, and pan into the view you want to print and then go to **File**, **Export**, select **2D Graphic**, and for File type, choose either .tiff or .jpeg. Explanations of each of these file extensions is listed in Chapter 4 of this book. In Photoshop, go to **File**, then **Open**, and find the document.

Images in InDesign

Projects should be converted to .jpeg for inclusion in an InDesign document. As described above, once AutoCAD and SketchUp projects get converted to a format that can be read in Photoshop, the image can then be saved as a .jpeg. Never copy and paste an image from Photoshop to InDesign: the image loses resolution and it becomes unmanageable when using the Links Palette. Always save the image as a .jpeg. Create a picture frame box and go to **File** and then Place to locate and insert an image.

Student Project Examples from Ashley Calabria and Jose R. Buitrago (Graduate and Undergraduate Levels)

The following images (figures 6-1 to 6-21) were primarily created by the authors' students from an introduction to computer graphics course, a portfolio development course that utilizes Photoshop for imagery and rendering and other course projects that were based on skills learned in the aforementioned classes. Their imaginative resourcefulness was the authors' inspiration for this textbook.

The results of these program exchanges can be seen in their examples. To our students, our sincerest gratitude.

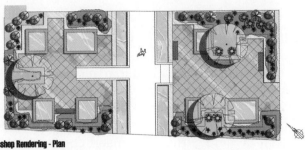

Photoshop Rendering - Plan
Scale 1 : 40

SketchUp - View From Street   **SketchUp - Inside View**

FIGURE 6–1 AutoCAD-drafted and brought into both Photoshop for rendering and SketchUp for 3D modeling. By Birt Garner for Design Communication II—Introduction to Computer Graphics (class review): Professor Calabria.

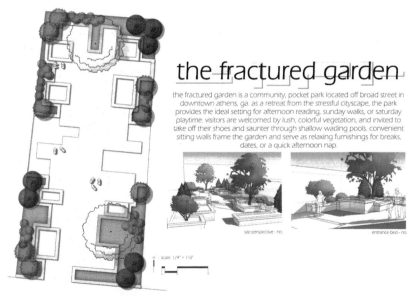

FIGURE 6–2 AutoCAD-drafted and brought into both Photoshop for rendering and SketchUp for 3D modeling. By Thomas Brown for Introduction to Computer Graphics: Professor Calabria.

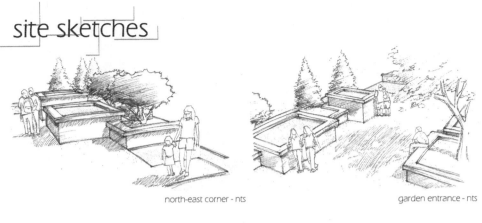

site sketches

north-east corner - nts   garden entrance - nts

the fractured garden

FIGURE 6–3 Views from SketchUp were printed out, hand drawn, and scanned back in for sheet layout. By Thomas Brown for Introduction to Computer Graphics: Professor Calabria.

Perspective view from the street   NTS

Perspective view from north-east corner   NTS

N

Master Plan
Scale 1 : 50

FIGURE 6–4 AutoCAD-drafted and brought into both Photoshop for rendering and SketchUp for 3D modeling. By Heath Tucker for Introduction to Computer Graphics: Professor Calabria.

SketchUp
Views

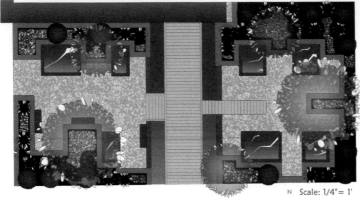

Photoshop
Plan

N Scale: 1/4"= 1'

FIGURE 6–5 AutoCAD-drafted and brought into both Photoshop for rendering and SketchUp for 3D modeling. By Jonathan Warner for Introduction to Computer Graphics: Professor Calabria.

CENTRAL STREET

FIGURE 6–6 3D SketchUp model. By Chris Lazarek, senior project advised by M. Akers and G. Coyle.

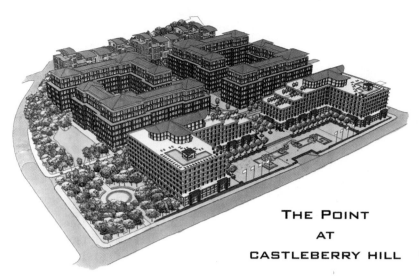

THE POINT
AT
CASTLEBERRY HILL

FIGURE 6–7 3D SketchUp model. By Chris Lazarek, senior project advised by M. Akers and G. Coyle.

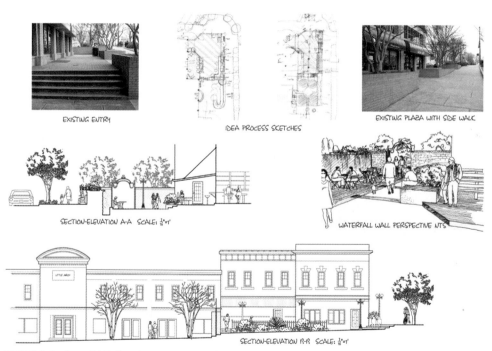

EXISTING ENTRY

IDEA PROCESS SKETCHES

EXISTING PLAZA WITH SIDE WALK

SECTION-ELEVATION A-A SCALE: ¾"=1'

WATERFALL WALL PERSPECTIVE NTS

SECTION-ELEVATION B-B SCALE: ¾"=1'

FIGURE 6–8 AutoCAD drawings, concept sketches, and hand sketches based on SketchUp modeling. By Zhen Feng designed in LAND 6040 Community and Place with Professor David Spooner and recreated for Introduction to Computer Graphics: Professor Calabria.

FIGURE 6–9 Hand drawn and Photoshop rendered using a tablet and stylus. By Curtis Alter designed in Beginning Design for Marguerite Koepke and rendered in Photoshop for Portfolio Development: Professor Calabria.

FIGURE 6–10 Before and after the Photoshop digital imagery. By Christine Donhardt for Introduction to Computer Graphics: Professor Calabria.

FIGURE 6–11   Photoshop digital imagery. By Audrey Thain for Portfolio Development: Professor Calabria.

FIGURE 6–12 Photoshop digital imagery. By Alison Peckett for Portfolio Development: Professor Calabria.

FIGURE 6–13 AutoCAD-drafted and rendered in Photoshop, SketchUp models and Photoshop digital imagery. By Christine Donhardt for First Year Design Studio with Professor Brian LaHaie.

FIGURE 6–14 AutoCAD drawing rendered in Photoshop as color pencil by Catherine Hawkins, Introduction to Computer Graphics: Professor Buitrago.

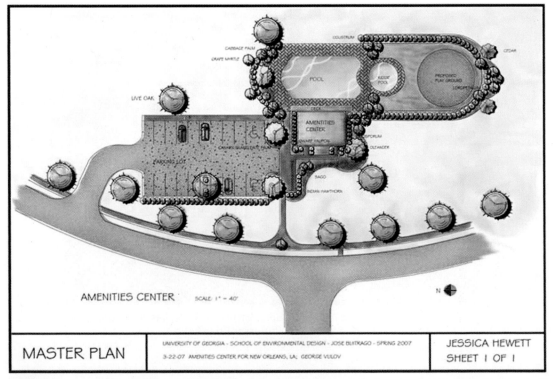

FIGURE 6–15 AutoCAD drawing rendered in Photoshop as air brush technique by Jessica Hewett, Introduction to Computer Graphics Course: Professor Buitrago.

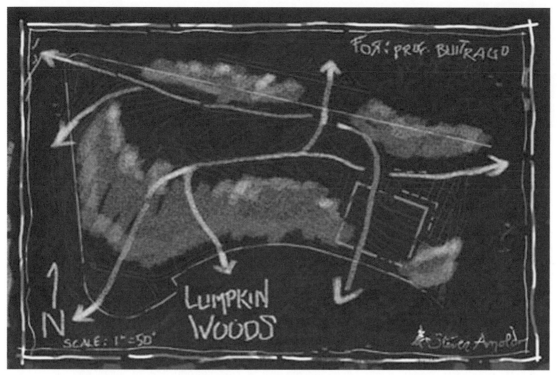

FIGURE 6–16 AutoCAD drawing rendered in Photoshop as color chalk technique by Steve B. Arnold, Advanced Computer Graphics Course: Professor Buitrago.

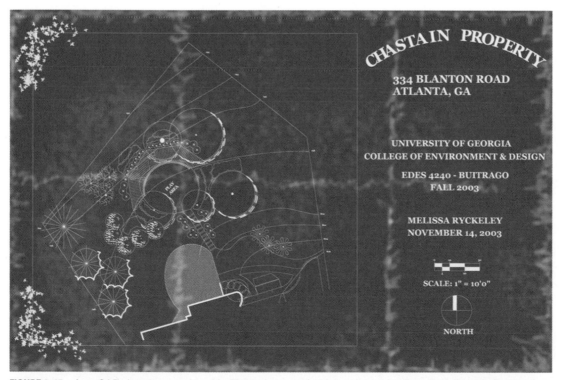

FIGURE 6–17 AutoCAD drawing rendered in Photoshop as old blue print by Melissa Ryckely, Advanced Computer Graphics Course: Professor Buitrago.

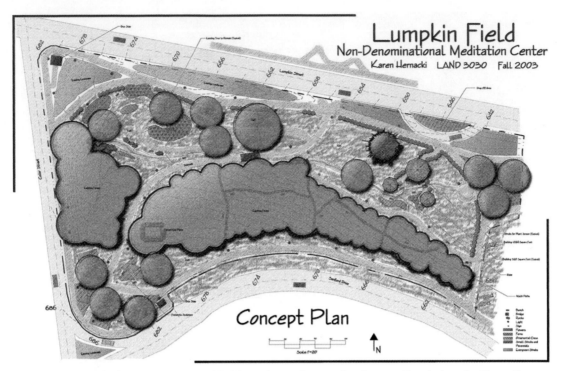

FIGURE 6–18 AutoCAD drawing rendered in Photoshop using a soft color pencil technique by Karen S. Hernacki, Advanced Computer Graphics Course: Professor Buitrago.

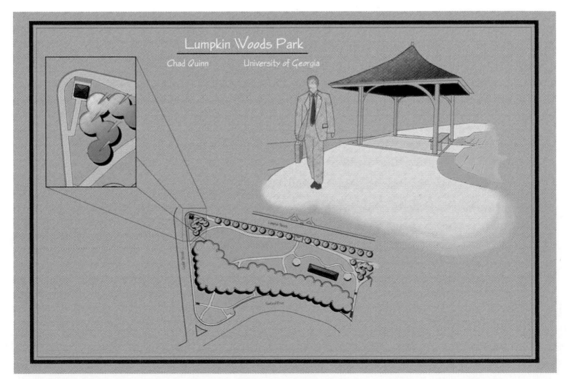

FIGURE 6–19 AutoCAD drawing rendered in Photoshop using a soft color pencil on sepia background look technique by Chad D. Quinn, Introduction to Computer Graphics Course: Professor Buitrago.

FIGURE 6–20 AutoCAD drawing rendered in Photoshop using a water color technique on texture background look technique by Cecily C. Williams, Introduction to Computer Graphics Course: Professor Buitrago.

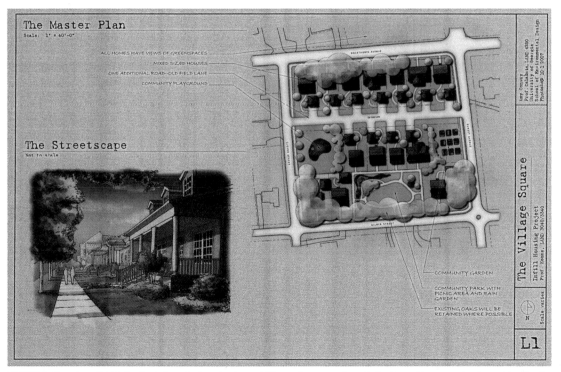

FIGURE 6–21 AutoCAD-drafted, hand-sketched, and Photoshop-rendered. By Amy Conway designed in Design Studio for Josh Koons and rendered in Photoshop for Portfolio Development: Professor Calabria.

Text Buttons

Chapter 1

All Programs	Hatch and Gradient	Page Setup
Alternate Units	Hatch Pattern Palette	Page Setup Manager
AutoDesk	Help	Plot
AutoCAD 2007	Image	Plot Setup
Block	Image Manager	Plotter Configuration Editor
Close	Index Color	Polar Tracking
Create New Dimension Style	Insert	Primary Units
Custom	Layer Properties Manager	Printers and Faxes
Dimension	Linear	Properties Palette
Dimension Style	Linetype	Raster Image
Dimension Style Manager	Lineweight	Select Color
Drafting Setting	Lines	Select Image File
Draw	Make	Select Linetype
Drawing Units	Modify	Set Current
Edit	Multileader	Set Current Layer
Express	My Dimension Style	Snap and Grid
External Reference	New	Styles
File	New Dimension Style	Symbols and Arrows
Fit	Object Snap	Text
Format	Override	Text Formatting

Text Style

Tolerance

Tools

Trim

View

Viewports

Window

Write Block

Chapter 2

2D Graphic

Add Scene

Animation

Camera

Components

Draw

Edit

Export

File

Freehand Tool

Hide

Import

Layers

Location

Materials

Model Info

Open

Print

Print Setup

Save As

Select Parallel Projection

Shadow Settings

Shadows

Styles

Text

Tools

Unhide

View

Window

Chapter 3

Adjustments

Adobe Acrobat Professional

Adobe PDF Document Properties

Adobe PDF Settings

Basic Brushes

Brightness/Contrast

Brush Editing

Brush Editor

Brush Option

Brush Type

Canvas Size

Color Picker

Copy

Custom Properties

Dry Media Brushes

Edit

File

Filter

Flatten Image

Format

Gradient Editor

Hue/Saturation

Image

Import PDF

Layer Box

Layer Editor

Layer Management

Layers

Layout

Merge Visible

Open

Paper Size

Paper/Quality

Paste

Plot

Plot File

Plot Scale

Plot Size

Plotter Configuration Editor

Preview

Print

Print/Save

RGB Color

Save

Save As

Save Adobe PDF

Select

Texture

Texturizer

Texturizer Editor

View

Wet Media Brushes

 Chapter 4

Adjustments

Brightness/Contrast

Color Balance

Copy

Duplicate Layer

Drop Shadow

Edit

File

Filter

Flatten Image

Flip Horizontal

Free Transform

General

History States

Hue/Saturation

Image

Image Size

Layer

Layer Styles

Layer Via Copy

Layer Via Cut

Levels

Load Selection

Open

Paste

Preferences

Save

Save As

Save for Web

Save Selection

Save Workspace

Select

Transform

Vanishing Point

Window

Workspace

Chapter 5

Document

Edit

File

File Types

Folder Options

New

New Document

Package

Preferences

Save

Save Workspace

Tools

Units and Increments

View

Window

Workspace

Chapter 6

2D Graphic

File

Open

Export

Import

Plot

Index

Note: Figures/illustrations, tables, and author's notes are indicated by f, t, and n respectively, following the page number.